Giuseppe Grioli (Ed.)

Proprietà di media e teoremi di confronto in fisica matematica

Lectures given at the
Centro Internazionale Matematico Estivo (C.I.M.E.),
held in Bressanone (Bolzano), Italy,
June 30- July 9, 1963

 Springer

C.I.M.E. Foundation
c/o Dipartimento di Matematica "U. Dini"
Viale Morgagni n. 67/a
50134 Firenze
Italy
cime@math.unifi.it

ISBN 978-3-642-11017-7   e-ISBN: 978-3-642-11018-4
DOI:10.1007/978-3-642-11018-4
Springer Heidelberg Dordrecht London New York

Printed on acid-free paper

Springer.com

CENTRO INTERNATIONALE MATEMATICO ESTIVO
(C.I.M.E)

Reprint of the 1st ed.- Bressanone, Italy, June 30-July 9, 1963

PRIOPRIETÀ DI MEDIA E TEOREMI DI CONFRONTO IN FISICA MATEMATICA

B. D. Coleman: On global and local forms of the second law of thermodynamics 1

J. Serrin: Comparison and averaging methods in mathematical physics 43

H. Ziegler: Thermodynamic aspects of continuum mechanics 133

C. Agostinelli: Un teorema di media sul flusso di energia nel moto di un fluido di alta conduttività elettrica in cui si genera un campo magnetico 165

Su alcuni teoremi di media in magnetofluidodinamica nel caso stazionario 171

D. Graffi: Principi di minimo e variazionali nel campo elettromagnetico 181

Teoremi di reciprocità nei fenomeni non stazionari 189

G. Grioli: Proprietà generali di media nella meccanica dei continui e loro applicazioni 201

Problemi di integrazione nella teoria dell'equilibrio elastico 217

CENTRO INTERNAZIONALE MATEMATICO ESTIVO

(C. I. M. E.)

B. D. COLEMAN

ON GLOBAL AND LOCAL FORMS
OF THE SECOND LAW OF THERMODYNAMICS

ROMA - Istituto Matematico dell'Università

Preface

The mathematical methods used here were set forth in the following two
articles :

(1) "Thermodynamics of elastic materials with heat conduction and viscosity", B.D. Coleman and W. Noll, Archive for Rational Mechanics
and Analysis $\underline{13}$, 167-178 (1963).

(2) "Thermodynamics and departures from Fourier's Law of heat conduction", B.D. Coleman and V.J. Mizel, Archive for Rational Mechanics and Analysis $\underline{13}$, 245-261 (1963).

Parts of the present text have been taken, with alterations and elaborations,
from (1). These lectures are concerned, however, mainly with some new
research to be published shortly by B.D. Coleman and V.J. Mizel in an arti-
cle entitled "Existence of caloric equations of state in thermodynamics".

B. D. Coleman

Lecture I

§ 1. Introduction

The basic physical concepts of classical continuum mechanics are <u>body</u>, <u>configuration</u> of a body, and <u>force system</u> acting on a body. In a formal rational development of the subject, one first tries to state precisely what mathematical entities represent these physical concepts. In rough language, a body is regarded to be smooth manifold whose elements are the material points; a configuration is defined as a mapping of the body into a three-dimensional Euclidean space, and a force system is defined to be a vector-valued function defined for pairs of bodies.* Once these concepts are made precise one can proceed to the statement of <u>general principles</u>, such as the principle of objectivity or the law of balance of linear momentum, and to the statement of specific <u>constitutive assumptions</u>, such as the assertion that a force system can be resolved into body forces with a mass density and contact forces with a surface density, or the assertion that the contact forces at a material point depend on certain local properties of the configuration at the point. While the general principles are the same for all work in classical continuum mechanics, the constitutive assumptions vary with the application in mind and serve to define the <u>material</u> under consideration. When one has stated the mathematical nature of bodies, configurations and forces, and has laid down the ways in which these concepts occur in the general principles and the constitutive assumptions, then the properties of these concepts are fixed, and one can present rigorous arguments without recourse to "operational definitions" and other metaphysical paraphernalia, which may be of some use in deciding

---

* For more extensive discussions of the foundations of continuum mechanics see references [1] - [4] .

B. D. Coleman

on the applicability of a theory to a specific physical situation but seem to have no place in its mathematical development.

Albeit the problem of the formulation of a detailed list of axioms for mechanics still has, even for the experts, some troublesome open questions, we can still assume in these lectures that we have sufficient familiarity with continuum mechanics to use the basic concepts and principles of the subject without continual reference to such a list.

To discuss the thermodynamics of continua, it appears that to the concepts of continuum mechanics one must add five new basic concepts: these are temperature, specific internal energy[*], specific entropy[**], heat flux, and heat supply[***] (due to radiation). Once mechanics is axiomatized, it is easy to give the mathematical entities representing the thermodynamic concepts: temperature, specific internal energy, specific entropy and heat supply are scalar fields defined over the body, while heat flux is a vector field over the body. I believe that in presenting thermodynamics one should retain all the general priciples of mechanics but add to them two new principles: the first law of thermodynamics, i.e. the law of balance of energy[****], and the second law, which for continua takes the form of the Clausius-Duhem inequality[*****]. Of course in thermodynamics one must make constitutive assumptions which involve some of the new variables which the subject introduces. The main

---

[*] Sometimes called "internal energy density".

[**] Sometimes called "entropy density".

[***] Sometimes called "density of absorbed radiation".

[****] Cf. §§ 241 and 242 of [4] .

[*****] Cf. § 257 of [4] .

B.D. Coleman

purpose of these lectures will be to examine the restrictions which the second law places on constitutive assumptions.[*]

Generalizing some earlier work of Truesdell [6] , Truesdell and Toupin [4] have formulated the following principle of equipresence: "a variable present as an independent variable in one constitutive equation should be so present in all". In other words, one should start a theory by assuming that all causes contribute to all effects. If one suspects a certain separation of effects one should not assume it a priori but should rather prove that general physical principles or assumed material symmetries require the separation. In their qualitative explanation of their original formulation of this principle, Truesdell and Toupin emphasized the separation of effects due to the invariance requirements of material objectivity and symmetry. I at first found myself unable to believe in the usefulness of equipresence, but a study of the consequences of thermodynamics restrictions [7] on constitutive equations has changed my viewpoint.

Here we shall use equipresence and assume that an independent variable present in one constitutive equation is so present in all, unless its presence is in direct contradiction with the assumed symmetry of the material, the principle of material objectivity or the laws of thermodynamics.

One of the things which we shall do here is to show that it is possible to use equipresence to motivate the classical linear thery of viscous fluids with heat conduction, although a cursory examination of the constitutive equations of that theory can yield the specious conclusion that the theory does not allow every cause to contribute to every effect.

---

[*] This concept of the structure of thermodynamics is explained in more detail in [5] .

B. D. Coleman

On Notation

We shall use the <u>direct</u>, as distinguished from the <u>component</u>, tensor notation, denoting vectors and points in Euclidean space by boldface Latin minuscules and tensors by lightface Latin majuscules. Tensors of order higher than two will not occur. We shall denote the transpose of a tensor F by F^T . The tensor Q will be said to be orthogonal if $QQ^T = Q^TQ = I$, where I is a unit tensor. The symbol $\underset{\sim}{0}$ will always denote the zero vector, but 0 may denote either the scalar zero or the zero tensor.

B. D. Coleman

§ 2. Thermodynamic Processes

Consider a body consisting of material points X . A thermodynamic process for this body is described by eight functions of X and the time t, with physical interpretations as follows:

(1) The spatial <u>position</u> $\underset{\sim}{x} = \underset{\sim}{\chi}(X, t)$; here the function $\underset{\sim}{\chi}$, called the <u>deformation function</u>, describes a <u>motion</u> of the body.

(2) The symmetric <u>stress</u> tensor $T = T(X, t)$.

(3) The body force $\underset{\sim}{b} = \underset{\sim}{b}(X, t)$ per unit mass (exerted on the body by the external world).

(4) The specific <u>internal energy</u> $\mathcal{E} = \mathcal{E}(X, t)$.

(5) The <u>heat flux</u> vector $\underset{\sim}{q} = \underset{\sim}{q}(X, t)$.

(6) The <u>heat supply</u> $r = r(X, t)$ per unit mass and unit time (absorbed by the material and furnished by radiation from the external world).

(7) The specific <u>entropy</u> $\eta = \eta(X, t)$.

(8) The local <u>temperature</u> $\theta = \theta(X, t)$, which is assumed to be always positive, $\theta > 0$.

We say that such a set of eight functions is a <u>thermodynamic process</u> [5] if the following two consevation laws[*] are satisfied not only for the body but for each of its parts $\mathcal{B}$:

(A) The law of balance of linear momentum:

$$(2.1) \qquad \int_{\mathcal{B}} \underset{\sim}{\ddot{x}}\, dm = \int_{\mathcal{B}} \underset{\sim}{b}\, dm + \int_{\partial\mathcal{B}} T\underset{\sim}{n}\, ds .$$

---

[*] A thorough discussion of these conservation laws is given in [4] , §§ 196--205, 240, 241.

B.D. Coleman

(B) The law of balance of energy

$$(2.2) \quad \frac{1}{2} \frac{d}{dt} \int_{\mathcal{B}} \dot{\underset{\sim}{x}} \, \ddot{\underset{\sim}{x}} \, dm + \int_{\mathcal{B}} \dot{\varepsilon} \, dm = \int_{\mathcal{B}} (\dot{\underset{\sim}{x}} \cdot b + r) dm + \int_{\partial \mathcal{B}} (\dot{\underset{\sim}{x}} \cdot T\underset{\sim}{n} - \underset{\sim}{q} \cdot \underset{\sim}{n}) ds.$$

In (2.1) and (2.2), dm denotes the element of mass in the body, $\partial \mathcal{B}$ the surface of $\mathcal{B}$, ds the element of surface area in the configuration at time t, and $\underset{\sim}{n}$ the exterior unit normal vector to $\partial \mathcal{B}$ in the configuration at time t; a superimposed dot denotes the material time derivative, i. e. the derivative with respect to t keeping X fixed. $\mathcal{B}$ and $\underset{\sim}{\chi}$ are assumed to be such that the <u>region</u>, $\underset{\sim}{\chi}(\mathcal{B}, t)$, occupied by $\mathcal{B}$ is, for each t, the closure of a bounded open connected set possessing a piecewise smooth surface.

The assumed symmetry of the stress tensor T insures that the moment of momentum is automatically balanced. Couple stresses, body couples and other mechanical interactions not included in T or $\underset{\sim}{b}$ are assumed to be absent.

Under suitable smoothness assumptions the balance equations (2.1) and (2.2) in integral form are equivalent to the following two balance equations in differential form[*] :

$$(2.3) \qquad \qquad \mathrm{div}\, T - \rho \ddot{\underset{\sim}{x}} = -\rho \underset{\sim}{b},$$

$$(2.4) \qquad \qquad \mathrm{tr}\{TL\} - \mathrm{div}\, \underset{\sim}{q} - \rho \dot{\varepsilon} = -\rho r.$$

Here ρ denotes the mass density; L is the velocity gradient, i.e. $L = \mathrm{grad}\, \dot{\underset{\sim}{x}}$; tr is the trace operator; and the operators grad and div refer

---

[*] See the sections of [4] cited above.

B.D. Coleman

to <u>spatial</u> derivatives, i.e. the gradient and divergence with respect to $\underset{\sim}{x}$ keeping t fixed.

We note that in order to define a thermodynamic process it suffices to prescribe the six functions $\underset{\sim}{\chi}$, T, $\mathcal{E}$, $\underset{\sim}{q}$, η, and θ. The remaining functions $\underset{\sim}{b}$ and r are then determined by (2.3) and (2.4).

It is often convenient to identify the material point X with its position $\underset{\sim}{X}$ in a fixed reference configuration R and to write

$$(2.5a) \qquad \underset{\sim}{x} = \underset{\sim}{\chi}(\underset{\sim}{X}, t) .$$

The gradient F of $\underset{\sim}{\chi}(\underset{\sim}{X}, t)$ with respect to $\underset{\sim}{X}$, i.e.

$$(2.5b) \qquad F = F(\underset{\sim}{X}, t) = \nabla \underset{\sim}{\chi}(\underset{\sim}{X}, t) ,$$

is called the <u>deformation gradient</u> at $\underset{\sim}{X}$ (i.e. at X) relative to the configuration R. It is well known that

$$(2.6) \qquad \dot{F} = LF . \qquad \text{i.e.} \quad L = \dot{F}F^{-1} ,$$

We assume that $\underset{\sim}{\chi}(\underset{\sim}{X}, t)$ is always smoothly invertible in its first variable, i.e. that the inverse F^{-1} of F exits, or, equivalently, that $\det F \neq 0$. We consistently use the symbol ∇ to indicate a gradient in the reference configuration R, i.e. a gradient computed taking $\underset{\sim}{X}$ as the independent variable, whereas grad is used when the position $\underset{\sim}{x}$ in the present configuration is taken as the independent variable. For a scalar field over $\mathcal{B}$, such as θ, it is easily shown that

$$(2.7) \qquad \nabla \theta = F^{T} \text{ grad } \theta .$$

B. D. Coleman

Since grad θ occurs often in our subject, it is convenient to have a single symbol for this vector. Let use the abbreviation

$$(2.8) \qquad\qquad \mathbf{g} \equiv \operatorname{grad} \theta \ .$$

The mass density ϱ is determined by F through the equation

$$(2.9) \qquad\qquad \varrho = \frac{1}{|\det F|}\, \varrho_r$$

where ϱ_r is a positive number, constant in time and equal to the mass density in the reference configuration R, and $|\det F|$ is the absolute value of the determinant of F.

H. Ziegler

$$(3.5) \qquad \mathcal{U} = \int \varrho\, u\; dV \;,$$

where $\mathcal{U}$ denotes the specific intrinsic energy, dependent on the mechanical state of the element, i.e., on its deformation, and on the temperature. The influx of heat into the volume V is

$$(3.6) \qquad \mathcal{Z}^{(u)} = - \int q_k\, \nu_k\; dS \;,$$

where the vector q_k denotes the heat flux. Starting from (1.2) and observing that, in a continuum, the energy of an element is composed of its kinetic and intrinsic energies, we state the <u>first fundamental theorem</u> for the volume V in the following form :

The material rate of increase of the sum of the kinetic and intrinsic energies in equal to the rate of work of the exterior forces plus the heat influx.

The analytical form of this statement is

$$
\begin{aligned}
(3.7) \qquad \int \varrho (v_k \dot{v}_k + \dot{u})\, dV &= \int \varrho\, f_k v_k\, dV + \int (\sigma_{kl} v_k - q_l)\, \nu_l\; dS \\
&= \int (\varrho\, f_k v_k + \sigma_{kl} v_{k,l} + \sigma_{kl,l} v_k - q_{l,l})\, dV
\end{aligned}
$$

On account of (3.4) and the symmetry of σ_{kl} (3.7) reduces to

$$(3.8) \qquad \int \varrho\, \dot{u}\; dV = \int (\sigma_{kl} v_{kl} - q_{k,k})\, dV \;,$$

where

$$(3.9) \qquad v_{kl} = \frac{1}{2}\, (v_{k,l} + v_{l,k})$$

B. D. Coleman

Lecture II

§ 3. Admissible Processes and Constitutive Assumptions

We assume that the __material__ at the point X is characterized by four __functions__ $\hat{\varepsilon}_{(X)}, \hat{\eta}_{(X)}, \hat{T}_{(X)}, \hat{q}_{(X)}$ which we call __response functions__ and which give ε, η, T, q at X when $\theta, g, F, \dot{F}$ are known at X :

$$(3.1) \qquad \varepsilon = \hat{\varepsilon}_{(X)}(\theta, g, F, \dot{F}) ,$$

$$(3.2) \qquad \eta = \hat{\eta}_{(X)}(\theta, g, F, \dot{F}) ,$$

$$(3.3) \qquad T = \hat{T}_{(X)}(\theta, g, F, \dot{F}) ,$$

$$(3.4) \qquad q = \hat{q}_{(X)}(\theta, g, F, \dot{F}) .$$

We say that a thermodynamic process in $\mathcal{B}$ is an __admissible thermodynamic process__ [5] if it is compatible with the constitutive equations (3.1)-(3.4) .

In dealing with response functions it is often important to distinguish between them and their values. Here a symbol with a superimposed $\wedge, \sim, -,$ or $=$ always denotes a __function__. Since, for a given process, the values of F and $\dot{F}$ must depend on the choice of the reference configuration R , the response functions $\hat{\varepsilon}_{(X)}, \hat{\eta}_{(X)}, \hat{T}_{(X)}, \hat{q}_{(X)}$ will depend on R . As the notation of (3.1)-(3.4) indicates, in general, these functions can also de-

B. D. Coleman

pend on the material point X . If there exists a reference configuration R^* of $\mathcal{B}$ which makes $\hat{\varepsilon}_{(X)},\ \hat{\eta}_{(X)},\ \hat{T}_{(X)},\ \hat{q}_{(X)}$ independent of X for all X in $\mathcal{B}$, then we say that $\mathcal{B}$ is <u>materially homogeneous</u> and that R^* is a homogeneous configuration of $\mathcal{B}$; if there is no such configuration R^* of $\mathcal{B}$, then $\mathcal{B}$ is <u>materially inhomogeneous</u>. For ease in writing, we shall drop the subscript (X) on response functions; however, <u>all the arguments we shall give here are valid equally for materially homogeneous and materially inhomogeneous bodies.</u>

In an admissible thermodynamic process, the arguments $\theta, g, F, \dot{F}$ and the values ε, η, T, q of the response functions $\hat{\varepsilon}, \hat{\eta}, \hat{T}, \hat{q}$ will, of course, depend on the time t . We assume that the functions $\hat{\varepsilon}, \hat{\eta}, \hat{T}, \hat{q}$ are themselves independent of t .

The constitutive equations considered here are not the most general imaginable; for example, they do not allow for all the long range memory effects covered in the purely mechanical theory of simple materials[*]. Our assumptions are, however, sufficiently general to cover many applications; in particular, they include as special cases the constitutive equations of the classical theories of thermoelastic phenomena and the hydrodynamics of viscous fluids with heat conduction. In contradistinction to the usual presentations of these classical theories, <u>we here, in Eqs.</u> (3.1)-(3.4), <u>start with constitutive assumptions that are compatible with the principle of equipresence.</u>

We do not lay down constitutive equations for body force density $\underset{\sim}{b}$ and the heat supply r due to absorbed radiation. <u>The quantities $\underset{\sim}{b}$ and r are regarded as assignable;</u> they can be assigned any values compatible with the balance equations (2.3) and (2.4) . Let us elaborate on the physical

---

[*] Cf. [2] & [8] .

B. D. Coleman

significance of this assumption. Let X be a material point in $\mathcal{B}$. In
the present theory we are following standard procedure and are ignoring mu-
tual body forces and self-radiation within $\mathcal{B}$. Here $\underset{\sim}{b}$ and r at X
depend not only on the "local state" $(\theta, \underset{\sim}{g}, \widetilde{F}, \overset{\bullet}{F})$ at X but also on the "external world", i.e. on the state of regions outside of $\mathcal{B}$. Our mathemati-
cal assumption that $\underset{\sim}{b}$ and r are assignable has the physical meaning
that we suppose that for each local state at X one can adjust the conditions
outside of $\mathcal{B}$ so that r and $\underset{\sim}{b}$ take on arbitrary values compatible
with balance of momentum and energy. That an experimenter might prefer
to fix the outside conditions and thus lose freedom in assigning thermodynamic fields should not affect our proofs: the theorist can consider processes
which the experimenter finds difficult to realize, provided only that they are
not <u>impossible</u> to realize.

We assume that for any fixed set of values of $\underset{\sim}{g}, F, \overset{\bullet}{F}$ the function $\hat{\mathcal{E}}$
is smoothly invertible in its first variable θ ; i.e.,

$$(3.5) \qquad \frac{\partial \hat{\mathcal{E}}}{\partial \theta} (\theta, \underset{\sim}{g}, F, \overset{\bullet}{F}) \neq 0.$$

This implies that there exist functions $\widetilde{\theta}, \widetilde{\eta}, \widetilde{T}, \underset{\sim}{\widetilde{q}}$, also called response
functions, which can be used to rewrite (3.1)-(3.4) in the forms

$$(3.6) \qquad \theta = \widetilde{\theta} (\mathcal{E}, \underset{\sim}{g}, F, \overset{\bullet}{F})$$

$$(3.7) \qquad \eta = \widetilde{\eta} (\mathcal{E}, \underset{\sim}{g}, F, \overset{\bullet}{F})$$

$$(3.8) \qquad T = \widetilde{T} (\mathcal{E}, \underset{\sim}{g}, F, \overset{\bullet}{F})$$

$$(3.9) \qquad \underset{\sim}{q} = \widetilde{q}(\mathcal{E}, \underset{\sim}{g}, F, \overset{\bullet}{F})$$

B. D. Coleman

For each set of the quantities $g, F, \dot{F}$, the function $\tilde{\theta}(\cdot, g, F, \dot{F})$ is the inverse function of $\hat{\mathcal{E}}(\cdot, g, F, \dot{F})$, and $\tilde{\eta}$ is defined by

$$(3.10) \qquad \tilde{\eta}(\mathcal{E}, g, F, \dot{F}) \equiv \hat{\eta}(\tilde{\theta}(\mathcal{E}, g, F, \dot{F}), g, F, \dot{F}) .$$

$\tilde{T}$ and $\tilde{q}$ are defined by formulae analogous to (3.10).

To every choice of the deformation function χ and the temperature distribution θ, as functions of X and t, there corresponds a unique admissible thermodynamic process in $\mathcal{B}$. For, when $\chi(X, t)$ and $\theta(X, t)$ are known for all X and t, clearly $F, \dot{F}$, and θ are determined throughout $\mathcal{B}$. The constitutive equations (3.1)-(3.4) then determine $\mathcal{E}$, η, T, and q throughout $\mathcal{B}$. Once the fields $\chi, T, \mathcal{E}, q$, and θ are known, r and b are determined by the balance laws (2.3) and (2.4).

Let $\varphi(t)$ be any time-dependent positive scalar; $a(t)$ any time-dependent vector; $A(t)$ any time-dependent invertible tensor; and Y any material point of $\mathcal{B}$ whose spatial position in the reference configuration R is Y. We can always construct at least one admissible thermodynamic process in $\mathcal{B}$ such that $\theta(X, t)$, $g(X, t)$, $F(X, t)$ have, respectively, the values $\varphi(t)$, $a(t)$, $A(t)$ at $X = Y$. An example of such a process is the one determined by the following deformation function and temperature distribution:

$$(3.11a) \qquad x = \chi(X, t) = Y + A(t)\left[X - Y\right],$$

$$(3.11b) \qquad \theta = \theta(X, t) = \varphi(t) + \left[A^T(t)a(t)\right] \cdot \left[X - Y\right],$$

B. D. Coleman

i.e.,

$$(3.11b') \qquad \theta = \theta(\underset{\sim}{x}, t) = \underset{\sim}{\alpha}(t) + \underset{\sim}{a}(t) \cdot \left[\underset{\sim}{x} - \underset{\sim}{\chi}\right]$$

where $\underset{\sim}{x} = \underset{\sim}{\tilde{\chi}}(\underset{\sim}{Y}, t) = \underset{\sim}{Y}$. Thus, at a given time t , we can arbitrarily specify ont only $\theta, \underset{\sim}{g}$ and F but also their time derivatives $\dot{\theta}, \dot{\underset{\sim}{g}}, \dot{F}, \ddot{F}$, etc. at a point $\underset{\sim}{Y}$ and be sure that there exists at least one admissible thermodynamic process corresponding to this choice. Furthermore, it follows from this, (3.1) and (3.5) that $\mathcal{E}, \underset{\sim}{g}, F,$ and the time-derivatives $\dot{\mathcal{E}}, \dot{\underset{\sim}{g}}, \dot{F}, \ddot{F}$ also form a set of quantities which can be chosen independently at one fixed point and time.

B. D. Coleman

Lecture III

§ 4. The Clausius-Duhem Inequality and Its Consequences

We regard q/θ to be the vectorial <u>flux of entropy</u> due to heat flow and r/θ to be a scalar <u>supply</u> of entropy from radiation. In other words, for each process we define the <u>rate of production of entropy</u> in the part $\mathcal{B}$ to be

$$(4.1) \qquad \Gamma = \frac{d}{dt} \int_{\mathcal{B}} \dot{\eta}\, dm - \int_{\mathcal{B}} \frac{r}{\theta}\, dm + \int_{\partial\mathcal{B}} \frac{1}{\theta}\, q \cdot n\, ds$$

where dm is the element of mass in $\mathcal{B}$, n the exterior unit normal to the surface $\partial\mathcal{B}$ of $\mathcal{B}$, and ds the element of surface area in the configuration at time t. Under appropriate smoothness assumptions we can write

$$(4.2) \qquad \Gamma = \int_{\mathcal{B}} \gamma\, dm$$

where

$$(4.3) \qquad \gamma = \dot{\eta} - r/\theta + \rho^{-1} \operatorname{div} q/\theta$$

$$= \dot{\eta} - \frac{r}{\theta} + \frac{1}{\rho\theta} \operatorname{div} q - \frac{1}{\rho\theta^2}\, q \cdot g$$

is the specific rate of production of entropy.

One way [5] of giving the Second Law of Thermodynamics a precise mathematical meaning is to lay down the following postulate.

B. D. Coleman

Postulate: For every admissible thermodynamic process in a body, the following inequality must hold for all t and all parts $\mathcal{B}$ of the body:

$$(4.4) \qquad \Gamma \geq 0 .$$

The inequality (4.4) is called the Clausius-Duhem inequality. Our postulate places restrictions on constitutive equations of the type (3.1)-(3.4), $\big[$ or (3.6)-(3.9)$\big]$. We now attempt to find necessary and sfficient set of such restrictions.

In order that (4.4) holds for all parts $\mathcal{B}$ of a body, it is necessary and sufficient that

$$(4.5) \qquad \gamma \geq 0$$

at all material points X of the body.

For each thermodynamic process, the energy balance equation (2.2) permits us to rewrite (4.3) as follows

$$(4.6) \qquad \gamma = \dot{\eta} - \frac{\dot{\varepsilon}}{\theta} + \frac{1}{\varrho\theta} \, \text{tr} \, \{TL\} - \frac{1}{\varrho\theta^2} \, \underset{\sim}{q} \cdot \underset{\sim}{g}$$

In an admissible process $\underset{\sim}{q}$ and   T   must be given by (3.8) and (3.9), and $\dot{\eta}$ must be given by

$$(4.7) \qquad \dot{\eta} = \eta_\varepsilon \dot{\varepsilon} + \underset{\sim}{\eta}_g \cdot \underset{\sim}{g} + \text{tr} \, \{\eta_F \dot{F}\} + \text{tr} \, \{\eta_{\dot{F}} \ddot{F}\} ,$$

where η_ε is the (scalar) value of $\dfrac{\partial \tilde{\eta}}{\partial \varepsilon}$; η_g is the (vector)value   of the gradient of the function $\underset{\sim}{\tilde{\eta}}$ with respect to its second variable $\underset{\sim}{g}$; while η_F and $\eta_{\dot{F}}$ are, respectively, the (tensor) values of gradients of

$\tilde{\eta}$ with respect to its third variable F and its fourth variable $\dot{F}$. It follows from (4.6), (4.7), and (2.5) that

$$(4.8) \qquad \gamma = \eta_g \cdot \dot{g} + \mathrm{tr}\left\{\eta_{\dot{F}}\ddot{F}\right\} + (\eta_\varepsilon - \tfrac{1}{\theta})\dot{\varepsilon} + \mathrm{tr}\left\{\eta_F \dot{F}\right\} +$$

$$+ \mathrm{tr}\left\{\frac{1}{\rho\theta}\, T\dot{F}F^{-1}\right\} - \frac{1}{\rho\theta^2}\, q\cdot g \qquad .$$

On looking at (4.8), (3.6)-(3.9), and (2.7) we see that γ depends on only the values of the seven quantities, ε , g , F , $\dot{\varepsilon}$, $\dot{g}$, $\dot{F}$, $\ddot{F}$ at X and t . According to the remarks made at the end of Section 3 , these seven quantities can be independently and arbitrarily chosen at X and t , and there will always exist an admissible thermodynamic process corresponding to the choice. Our postulate (4.4) is equivalent to the assertion that γ be $\geqslant 0$ for all such choices.

To find the <u>necessary</u> conditions for the validity of our postulate first observe that (4.8) can be written in the form

$$(4.9) \qquad \gamma = \tilde{\eta}_g (\varepsilon,\ g,\ F,\ \dot{F})\cdot\dot{g} + f(\varepsilon,\ g,\ F,\ \dot{\varepsilon},\ \dot{F},\ \ddot{F}) \qquad .$$

If we assign ε , g , F , $\dot{\varepsilon}$, $\dot{F}$, $\ddot{F}$ any fixed values, $f(\varepsilon,\ g,\ F,\ \dot{\varepsilon},\ \dot{F},\ \ddot{F})$ will be fixed at some finite value, say $\underline{a}$, and the postulate will require that

$$(4.10) \qquad \tilde{\eta}_g(\varepsilon,\ g,\ F,\ \dot{F})\cdot\dot{g} + a \geqslant 0$$

for <u>all</u> values of $\dot{g}$. But clearly this is possible only if

B.D. Coleman

$$(4.11) \qquad \tilde{\eta}_g(\mathcal{E}, g, F, \dot{F}) = \underset{\sim}{0} \ .$$

Futhermore, this equation must hold for all values of $\mathcal{E}$, g, F, $\dot{F}$;i.e., η in (3.7) cannot depend on g .

It follows from (4.11) that (4.8) can be written in the form

$$(4.12) \qquad \gamma = \mathrm{tr}\left\{\tilde{\eta}_{\dot{F}}(\mathcal{E}, F, \dot{F})\ddot{F}\right\} + \ell(\mathcal{E}, g, F, \dot{\mathcal{E}}, \dot{F})$$

Hence for any fixed values of $\mathcal{E}$, g, $\dot{\mathcal{E}}$, F, $\dot{F}$, the postulate (4.4) requires that

$$(4.13) \qquad \mathrm{tr}\left\{\tilde{\eta}_{\dot{F}}(\mathcal{E}, F, \dot{F})\ddot{F}\right\} + b \geqslant 0$$

where b is a finite number. The inequality (4.13) can hold for all choices of $\ddot{F}$ only if

$$(4.14) \qquad \tilde{\eta}_{\dot{F}}(\mathcal{E}, F, \dot{F}) = 0$$

where 0 is the zero tensor. Since (4.14) must hold for all values of $\mathcal{E}$, F, and $\dot{F}$, we have proved that our postulate requires that not only g , but also $\dot{F}$, must drop out of (3.7) , i.e. that η must be given by a function $\tilde{\eta}$ of $\mathcal{E}$ and F alone :

$$(4.15) \qquad \eta = \tilde{\eta}(\mathcal{E}, F) \ .$$

Of course, the function $\tilde{\eta}$ depends on the point X under consideration.

It follows from what we have done so far that

B. D. Coleman

$$(4.16) \qquad \gamma = \left(\tilde{\eta}_{\mathcal{E}}(\mathcal{E}, F) - \frac{1}{\tilde{\theta}(\mathcal{E}, g, F, \dot{F})} \right) \dot{\mathcal{E}} + h(\mathcal{E}, g, F, \dot{F})$$

and a now familiar argument yields the result that if (4.4) is to hold for all admissible processes, then the coefficient of $\dot{\mathcal{E}}$ in (4.16) must be zero for each choice of $\mathcal{E}$, g, F, $\dot{F}$. Thus, the postulate implies that θ must also be given by a function $\tilde{\theta}$ of $\mathcal{E}$ and F alone,

$$(4.17) \qquad \qquad \theta = \tilde{\theta}(\mathcal{E}, F) \ ;$$

and that $\tilde{\theta}$ must be related to the function $\tilde{\eta}$ through the <u>temperature relation</u> :

$$(4.18) \qquad \qquad \tilde{\theta}(\mathcal{E}, F) = \left[\tilde{\eta}_{\mathcal{E}}(\mathcal{E}, F) \right]^{-1} \ .$$

Let $\tilde{T}^{(0)}(\mathcal{E}, F)$ give the <u>equilibrium stress</u> $T^{(0)}$ corresponding to $\mathcal{E}$ and F, i.e., the stress when the temperature gradient g and the velocity gradient $L = \dot{F} F^{-1}$ both vanish :

$$(4.19) \qquad \qquad \tilde{T}^{(0)}(\mathcal{E}, F) = \tilde{T}(\mathcal{E}, \underset{\sim}{0}, F, 0).$$

The value $T^{(e)}$ of the function $\tilde{T}^{(e)}$ defined by

$$(4.20) \qquad \tilde{T}^{(e)}(\mathcal{E}, g, F, \dot{F}) = \tilde{T}(\mathcal{E}, g, F, \dot{F}) - \tilde{T}^{(0)}(\mathcal{E}, F)$$

is called the <u>extra-stress</u> ; it is the contribution to T "caused" by the gradients of temperature and velocity.

Using (4.15), (4.17), (4.18), and (4.20) we can write (4.8) in the form :

$$(4.21) \quad \rho \, \tilde{\theta}(\mathcal{E}, F) \, \dot{\gamma} = \mathrm{tr} \left\{ \left[\rho \, \tilde{\theta}(\mathcal{E}, F) \tilde{\eta}_F(\mathcal{E}, F) + F^{-1} \tilde{T}^{(o)}(\mathcal{E}, F) \right] \dot{F} \right\}$$

$$+ \, \mathrm{tr} \left\{ \tilde{T}^{(e)}(\mathcal{E}, g, F, \dot{F}) \dot{F} F^{-1} \right\} - \frac{1}{\tilde{\theta}(\mathcal{E}, F)} \, \tilde{q}(\mathcal{E}, g, F, \dot{F}) \cdot g$$

In writing (4.21) we have made use of the fact that the trace is a linear function with the property that $\mathrm{tr}\left\{ T \dot{F} F^{-1} \right\} = \mathrm{tr}\left\{ F^{-1} T \dot{F} \right\}$. Assuming that $\tilde{T}$ is continuous in its fourth variable $\dot{F}$ at $\dot{F} = 0$, the following holds:

$$(4.22) \qquad \tilde{T}(\mathcal{E}, g, F, \alpha \dot{F}) = \tilde{T}(\mathcal{E}, g, F, 0) + o(1) ,$$

where α is a real number and $o(1)$ is such that for fixed values of $\mathcal{E}$, g, F, $\dot{F}$,

$$\lim_{\alpha \to 0} o(1) = 0 .$$

It follows from (4.22) and the definitions (4.19), (4.20) that

$$(4.23) \qquad \tilde{T}^{(e)}(\mathcal{E}, \underset{\sim}{0}, F, \alpha \dot{F}) = o(1) .$$

In (4.21) let us now put $g = \underset{\sim}{0}$ and replace $\dot{F}$ by $\alpha \dot{F}$ we find, using (4.23), that our postulate requires that

$$(4.24) \quad \rho \, \tilde{\theta}(F, \mathcal{E}) \, \dot{\gamma} = \alpha \, \mathrm{tr} \left\{ \left[\rho \, \tilde{\theta}(\mathcal{E}, F) \tilde{\eta}_F(\mathcal{E}, F) + F^{-1} \tilde{T}^{(o)}(\mathcal{E}, F) \right] \dot{F} \right\} +$$

$$+ \, o(\alpha) \geqslant 0 ,$$

where

B. D. Coleman

$$\lim_{\alpha \to 0} o(\alpha)/\alpha = 0 \ .$$

Equation (4.24) must hold for all values of ε, F, $\dot{F}$, and α. On considering the behavior of (4.24) for small values of α, we conclude that the coefficient of α must be zero; i.e., for each value of the pair ε, F, we must have

$$(4.25) \qquad \mathrm{tr}\left\{\left[\rho\tilde{\theta}(\varepsilon,F)\tilde{\eta}_F(\varepsilon,F) + F^{-1}\tilde{T}^{(o)}(\varepsilon,F)\right]\dot{F}\right\} = 0$$

for all values of $\dot{F}$. but this is possible only if the **stress-relation**

$$(4.26) \qquad \tilde{T}^{(o)}(\varepsilon,F) = -\rho\tilde{\theta}(\varepsilon,F)F\tilde{\eta}_F(\varepsilon,F)$$

holds. Equations (4.18) and (4.26) tell us that the equilibrium stress defined by (4.19) is determined when the caloric equation of state (4.15) is known.

It follows from (4.26), (4.21), and (2.5) that

$$(4.27) \qquad \rho\tilde{\theta}(\varepsilon,F)\gamma = \mathrm{tr}\left\{\tilde{T}^{(e)}(\varepsilon,g,F,\dot{F})L\right\} - \frac{1}{\tilde{\theta}(\varepsilon,F)}\tilde{g}(\varepsilon,g,F,\dot{F})\cdot g \ .$$

It follows from (4.27) and our postulate that when $g = \underset{\sim}{0}$ we have the **mechanical dissipation inequality**

$$(4.28) \qquad \mathrm{tr}\left\{\tilde{T}^{(e)}(\varepsilon,\underset{\sim}{0},F,\dot{F})L\right\} \geqslant 0 \ ,$$

and when $L = 0$ we have the **heat conduction inequality**

$$(4.29) \qquad \tilde{q}(\varepsilon,g,F,0)\cdot g \leqslant 0 \ .$$

B. D. Coleman

We note, however, that in general a resolution of the inequality $\rho\,\theta\,\gamma \geq 0$ into a mechanical dissipation inequality valid for <u>non-zero</u> g and a heat conduction inequality valid for <u>non-zero</u> L is not valid. It would be valid if the following conditions were met : that $\operatorname{tr}\left\{T^{(e)}L\right\}$ be independent of g and that $q \cdot g$ be independent of L . These conditions do not follow from the assumptions we have made so far, although they hold when one adds certain special assumptions of linearity and symmetry.

It is clear, from (4.27), that if (4.15), (4.18), and (4.26) are assumed, the inequality

$$(4.30) \qquad \operatorname{tr}\left\{T^{(e)}D\right\} - \frac{1}{\theta}\, q \cdot g \geq 0 \,,$$

which we call the <u>general dissipation inequality</u>, is not only necessary, but also sufficient for our postulate. In writing (4.30) we have used the fact that since $T^{(e)}$ is symmetric, $\operatorname{tr}\left\{T^{(e)}L\right\} = \operatorname{tr}\left\{T^{(e)}D\right\}$ where D is the symmetric part of L .

We summarize in the following

<u>Theorem</u> : Consider a body $\mathcal{B}$ and assume only that constitutive equations of the general form (3.1) - (3.4) hold at each X in $\mathcal{B}$. Under this assumption, a necessary and sufficient condition that the inequality (4.4) hold for all admissible thermodynamic processes in $\mathcal{B}$ is that the following statements be true at each X in $\mathcal{B}$.

I. There exists a caloric equation of state (4.15).

II. The temperature is given by the relations (4.17) and (4.18).

III. The equilibrium stress, defined in (4.19), is given by the stress-relation (4.26).

IV. The functions $\widetilde{T}^{(e)}$ of (4.20) and $\tilde{q}$ of (3.9) obey the general dis-

B. D. Coleman

sipation inequality (4.30) for all values of $\mathcal{E}$, F, grad θ, and D .
The inequality (4.4) implies the mechanical dissipation inequality $\mathrm{tr}\left\{T^{(e)}L\right\} \geqslant 0$
and the heat conduction inequality $q \cdot g \geqslant 0$ only in special circumstances,
such as those shown in (4.28) and (4.29) .

Addenda on Other Forms of the Temperature and Stress Relations

It follows from our assumption that $\theta > 0$ and the relation (4.18) that
the function $\tilde{\eta}$ is smoothly invertible in its first variable $\mathcal{E}$. Hence,
there exists a function $\bar{\mathcal{E}}$ (depending on X) giving $\mathcal{E}$ at X as a
function of η and F at X :

$$(4.31) \qquad \mathcal{E} = \bar{\mathcal{E}}(\eta, F).$$

Using this function, one can rewrite (4.17) and (4.18) in the form

$$(4.32) \qquad \theta = \bar{\theta}(\eta, F) = \mathcal{E}_\eta(\eta, F).$$

Equation (4.31) can also be used to express T and q as functions of
η , g, F and $\dot{F}$:

$$(4.33) \qquad T = \tilde{T}(\bar{\mathcal{E}}(F, \eta), g, F, \dot{F}) = \bar{T}(\eta, g, F, \dot{F})$$

$$(4.34) \qquad q = \tilde{q}(\bar{\mathcal{E}}(F, \eta), g, F, \dot{F}) = \bar{q}(\eta, g, F, \dot{F}).$$

and to express $T^{(o)}$ as a function of η and F

$$(4.35) \qquad T^{(o)} = \tilde{T}^{(o)}(\bar{\mathcal{E}}(F, \eta), F) = \bar{T}(\eta, F) = \bar{T}(\eta, 0, F, 0)$$

Using the chain rule we can cast (4.26) into the simpler form

$$(4.36) \qquad T^{(o)} = - \rho \, F \, \bar{\mathcal{E}}_F(\eta, F) \,.$$

It follows from our assumptions (3.1) and (3.5) that the function $\tilde{\theta}$ of (4.17) is smoothly invertible in $\mathcal{E}$, i.e., that $\mathcal{E}$ at X is given by a function $\bar{\mathcal{E}}$ of θ and F at X :

$$(4.37) \qquad \mathcal{E} = \bar{\bar{\mathcal{E}}}(\theta, F);$$

and, by (4.15) we have

$$(4.38) \qquad \eta = \tilde{\eta}\,(\bar{\bar{\mathcal{E}}}(\theta, F), F) = \bar{\bar{\eta}}\,(\theta, F)$$

Using (4.37) and (4.38) we can define a (Helmholtz) free-energy function $\bar{\bar{\psi}}$ in the usual way :

$$(4.39) \qquad \psi = \bar{\bar{\psi}}(\theta, F) = \bar{\bar{\mathcal{E}}}(\theta, F) - \theta \bar{\bar{\eta}}(\theta, F);$$

and it is not difficult to show that (4.18) and (4.26) $\big[$ or, equivalently, (4.32) and (4.36) $\big]$ yield

$$(4.40) \qquad \eta = - \bar{\bar{\psi}}_\theta(\theta, F)$$

$$(4.41) \qquad T^{(o)} = \rho \, F \, \bar{\bar{\psi}}_F(\theta, F).$$

B. D. Coleman

Lecture IV

§ 5. <u>Objectivity, Fluids</u>

The discussion of the present section will not require a separate notation for response functions and their values.

The theorem of Section 4 tells us that the second law of thermodynamics implies that constitutive equations of the type (3.1)-(3.4) reduce to the following equations

(5.1a)
$$\varepsilon = \mathcal{E}(\eta, F)$$

(5.1b)
$$\theta = \theta(\eta, F)$$

(5.1c)
$$T = T^{(o)}(\eta, F) + T^{(e)}(\eta, g, F, L)$$

(5.1d)
$$q = q(\eta, g, F, L) \quad .$$

In writing (5.1c) and (5.1d) we have used the fact that $L = \dot{F}F^{-1}$ and we have chosen η as an independent variable, in accord with (4.31)-(4.35), to obtain simpler formulae.

The reduced constitutive equations (5.1) must obey the principle of material objectivity [2], [3]. In rough language, this principle states that an admissible process remains admissible after a change of frame (or observer). We now consider the limitations placed on the Eqs.(5.1) by objectivity. The results will be intuitively plausible, and since the rigorous arguments which lead to them differ only in minor details from related proofs which in the last ten years have been frequently given in continuum physics,[*] I give here only

---

[*] Cf. [2] , [7] , [8] , [9] , [10] .

B. D. Coleman

a descriptive outline of a method which is in essence that used by Noll [2] , [8] in different contexts.

A change of frame is defined by a time-dependent orthogonal tensor Q. The scalars η , ε , and θ are unaffected by a change of frame, but F, $\dot{F}$, L, T and g = grad θ transform as follows :

$$F \rightarrow QF$$

$$\dot{F} \rightarrow \overline{\dot{QF}} = Q\dot{F} + \dot{Q}F$$

(5.2)
$$L = \dot{F}F^{-1} \rightarrow \overline{\dot{QF}}\,[QF]^{-1} = QLQ^T + \dot{Q}Q^T$$

$$T \rightarrow QTQ^T$$

$$q \rightarrow Qq$$

$$g \rightarrow Qg$$

The tensor $\dot{Q}Q^T$ is always skew.

The constitutive equations (5.1) are compatible with material objectivity if and only if the functions in (5.1) obey the following identities for each orthogonal tensor Q , each skew tensor W , and all values of ε , F, q, and L :

B.D.Coleman

$$\mathcal{E}(\eta, F) = \mathcal{E}(\eta, QF)$$

$$\theta(\eta, F) = \theta(\eta, QF)$$

(5.3)
$$QT^{(o)}(\eta, F)Q^T = T^{(o)}(\eta, QF)$$

$$QT^{(e)}(\eta, \underset{\sim}{g}, F, L)Q^T = T^{(e)}(\eta, Q\underset{\sim}{g}, QF, QLQ^T + W)$$

$$Q\underset{\sim}{q}(\eta, \underset{\sim}{g}, F, L) = \underset{\sim}{q}(\eta, Q\underset{\sim}{g}, QF, QLQ^T + W)$$

These identities can be used to derive the following reduced forms :

(5.4a)
$$\mathcal{E} = \mathcal{E}(\eta, C)$$

(5.4b)
$$\theta = \theta(\eta, C)$$

(5.4c)
$$F^T T F = \tilde{\tilde{T}}^{(o)}(\eta, C) + \tilde{\tilde{T}}^{(e)}(\eta, \nabla\theta, C, F^T D F)$$

(5.4d)
$$\underset{\sim}{q} = F\tilde{\tilde{\underset{\sim}{q}}}(\eta, \nabla\theta, C, F^T D F).$$

Here $C = F^T F$ is the left Cauchy-Green tensor ; $D = \frac{1}{2}(L + L^T)$ is again the stretching tensor; and $\nabla\theta$, which is related to $\underset{\sim}{g}$ by (2.6) , is the gradient of θ with respect to X . A material obeying (5.4) automatically obeys (5.1) and the identities (5.3) . In other words, the existence of the reduced forms (5.4) is not only a necessary, but also a sufficient, condition that the material under consideration be compatible with the principle of material objectivity.

The restrictions on the equations (5.1) caused by the symmetries which a material might possess will not be discussed here. A mathematical defini-

B. D. Coleman

tion of material symmetry (i.e. the "isotropy group") in materials obeying only slightly less general constitutive equations is given in reference[7]*.

We say that a material obeying (5.1) is a __fluid__ if the tensor F occurs in (5.1) only through $|\det F|$, or, by (2.8), only through the specific volume $v\ (= 1/\rho)$. It is not difficult to show that Eqs. (5.3) imply that for a fluid the Eqs. (5.1) must reduce to

$$(5.5a) \qquad \varepsilon = \varepsilon(\eta, \dot{v})$$

$$(5.5b) \qquad \theta = \theta(\eta, v)$$

$$(5.5c) \qquad T = T^{(o)}(\eta, v) + T^{(e)}(\eta, g, v, D)$$

$$(5.5d) \qquad q = q(\eta, g, v, D),$$

where $T^{(o)}$ is a hydrostatic pressure,

$$(5.6) \qquad T^{(o)}(\eta, v) = -p(\eta, v)I;$$

and $T^{(e)}$ and q are, for fixed η and v, isotropic functions of g and D, i.e., obey the following identities for every orthogonal tensor Q:

$$(5.7a) \qquad QT^{(e)}(\eta, g, v, D)Q^T = T^{(e)}(\eta, Qg, v, QDQ^T),$$

$$(5.7b) \qquad Qq(\eta, g, v, D) = q(\eta, Qg, v, QDQ^T).$$

---

* The definition given there is, in turn, analogous to that given in [2] and discussed in detail for elastic materials in [10].

B. D. Coleman

It follows from the definition (4.20) that

$$(5.8) \qquad T^{(e)}(\eta, \underset{\sim}{0}, \boldsymbol{v}, 0) = 0 .$$

For a fluid, Eq. (4.36) reduces to the following familiar expression for the equilibrium pressure function p in (5.6)

$$(5.9a) \qquad p(\eta, \boldsymbol{v}) = \varepsilon_{\boldsymbol{v}}(\eta, \boldsymbol{v}) ;$$

and (4.32) becomes

$$(5.9b) \qquad \theta(\eta, \boldsymbol{v}) = \varepsilon_{\eta}(\eta, \boldsymbol{v})$$

Let us return to the identities (5.7) . Representation theorems for such tensor-valued and vector-valued isotropic functions exist [*], but there is no need for us to state them in full generality here. Some special consequences of the identities of (5.7) may, however, be of interest. If in (5.7) we put $Q = -I$, then we obtain the identities

$$(5.10a) \qquad T^{(e)}(\eta, g, \boldsymbol{v}, D) = T^{(e)}(\eta, -g, \boldsymbol{v}, D)$$

$$(5.10b) \qquad -q(\eta, g, \boldsymbol{v}, D) = q(\eta, -g, \boldsymbol{v}, D) .$$

---

[*] For (5.7b) one can use directly the representation theorem for isotropic vector-valued functions of a vector and a symmetric tensor, proved by Pipkin and Rivlin [11] in a different context. In (5.7a) one can replace g by $g \otimes g$ and then use the representation theorems of Rivlin and Ericksen [12] for symmetric-tensor-valued functions of two symmetric tensors.

B. D. Coleman

Thus, for any fixed values of η, v, and D, $T^{(e)}$ must be an even function, and q an odd function, of g . In particular, we have

$$(5.11) \qquad q(\eta, \underline{0}, v, D) = 0 ;$$

i.e. in a fluid, regardless of the motion, there can be no heat flux under zero temperature gradient[*].

We now assume that $T^{(e)}$ and q , as functions of $\underline{D}$ and g , are differentiable at $D = 0, g = \underline{0}$ and consider approximation formulae for $T^{(e)}$ and q for small D and g . Since D and g are independently variable, and of different dimension, there is no intrinsecally preferred way of making precise the concept of a "first-order term in D __and__ g". It appears to me that the physicists' usual concept of a "linearized theory" corresponds to considering the space of vectors $D \oplus g$, using the "natural" norm $\| \ \|$ of that space,

$$(5.12) \qquad \| D \oplus g \| = \sqrt{\mathrm{tr}(D^2) + g \cdot g} \quad ,$$

and saying that an approximation is "complete to first-order" if it includes all terms $\geqslant o(\| D \oplus g \|)$. Now, using smoothness and known representation theorems for isotropic functions one can prove that Eqs. (5.7) imply

$$(5.13a) \qquad T^{(e)} = 2\mu D + \lambda (\mathrm{tr}\, D)I + o(\| D \oplus g \|)$$

---

[*] The present argument can be generalized to yield the same result for any material with the central inversion, -I, in its symmetry group. Note that this argument does not use the general dissipation inequality (4.30); cf. [5] and [7] .

B.D. Coleman

$$(5.13b) \qquad \underset{\sim}{q} = -k\underset{\sim}{g} + o(\| D \oplus \underset{\sim}{g} \|).$$

The scalars λ , μ , and k in (5.13) are functions of η and υ (or θ and υ) alone. We notice that to within terms $o(\| D \oplus \underset{\sim}{g} \|)$, $T^{(e)}$ is independent of $\underset{\sim}{g}$, and $\underset{\sim}{q}$ is independent of D . This is an obvious consequence of Eqs. (5.10) . Since (5.13) holds if (5.12) be replaced by

$$(5.12)' \qquad \| D \oplus \underset{\sim}{g} \| = \sqrt{\zeta_1 \mathrm{tr} D^2 + \zeta_2 \underset{\sim}{g} \cdot \underset{\sim}{g}}$$

where ζ_1 and ζ_2 are any two positive constants, the Eqs. (5.13) are invariant under changes of units, albeit they do not appear so at first glance.

The constitutive equations of the classical linear theory of viscous fluids with heat conduction are (5.5a)

$$(I) \qquad \mathcal{E} = \mathcal{E}(\eta, \upsilon) \quad ;$$

(5.6) and (5.9a)

$$(II) \qquad T^{(o)} = -p(\eta, \upsilon)I , \qquad p(\eta, \upsilon) = \mathcal{E}_\upsilon(\eta, \upsilon) ,$$

(5.5b) and (5.9b)

$$(III) \qquad \theta = \theta(\eta, \upsilon) = \mathcal{E}_\eta(\eta, \upsilon) ,$$

and the equations obtained by striking out the terms $o(\| D \oplus \underset{\sim}{g} \|)$ in Eqs. (5.13) :

B. D. Coleman

(IV)
$$T^{(e)} = 2\mu D + \lambda \, (tr\ D)I \, ,$$

(V)
$$q = kg$$

When terms $o(\|D \oplus g\|)$ are omitted from Eqs. (5.15) the general dissipation inequality (4.30) requires that both mechanical dissipation inequality

(5.14a)
$$tr\left\{T^{(e)}L\right\} \geqslant 0$$

and the heat conduction inequality

(5.14b)
$$q \cdot g \leq 0$$

hold. Given that $T^{(e)}$ has the form (IV), a necessary and sufficient condition that (5.14a) hold for all L is that both

(VI)
$$\mu \geqslant 0, \quad \text{and} \quad 3\lambda + 2\mu \geqslant 0 \, .$$

Of course, when q has the form (V), (5.14b) holds if and only if

(VII)
$$k \geqslant 0 \quad .$$

The inequalities (VI) and (VII) are just as basic to classical fluid dynamics as the equations (I)-(V).

Thus using general physical principles and starting from our constitutive equations (3.1)-(3.4) which reflect equipresence, we can derive the constitutive assumptions of classical fluid dynamics by adding only two specializing assertions : (1) that (3.1)-(3.4) do describe a fluid, i.e., that F enters

B.D. Coleman

only through $\left| \det F \right|$, and (2) that terms $o(\| D \oplus g \|)$ can be neglected in expressions for T and q .

- 34 -

$\S$ 6. <u>References</u>

[1] Noll, W: The foundations of classical mechanics in the light of recent advances in continuum mechanics. In: The Axiomatic Method with Special Reference to Geometry and Physics. Pp. 266-281. Amsterdam: North Holland Co. 1959.

[2] Noll, W: Arch. Rational Mech. Anal. $\underline{2}$, 197 (1958).

[3] Noll, W.: La mécanique classique, basée sur un axiome d'objectivité. A paper read at the Colloque Internationale sur la Méthode Axiomatique in Mécanique Classique et Moderne, Paris, 1959 (to be published by Gauthier-Villars, Paris).

[4] Truesdell, C., & R.A. Toupin: The Classical Field Theories. In: Encyclopedia of Physics, Vol. III/1, edited by S. Flügge. Berlin-Göttingen-Heidelberg : Springer 1960.

[5] Coleman, B.D., & W. Noll: Arch. Rational Mech. Anal. $\underline{13}$, 167 (1963).

[6] Truesdell, C.: J. Math. pures Appl. $\underline{30}$, 111(1951).

[7] Coleman, B.D., & V.J. Mizel : Arch. Rational Mech. Anal. $\underline{13}$, 245 (1963).

[8] Green, A.E., & R.S. Rivlin: Arch. Rational Mech. Anal. $\underline{1}$, 1(1957).

[9] Noll, W.: J. Rational Mech. Anal. $\underline{4}$, 3 (1955).

[10] Coleman, B.D., & W. Noll: Material symmetry and thermostatic inequalities in finite elastic deformations, Arch. Rational Mech. Anal. (in press).

[11] Pipkin, A.C., & R.S. Rivlin: J. Math. Phys. $\underline{1}$, 127 (1960).

[12] Rivlin, R.S., & J.L. Ericksen: J. Rational Mech. Anal. $\underline{4}$, 323 (1955).

CENTRO INTERNAZIONALE MATEMATICO ESTIVO

(C.I.M.E.)

43

J. SERRIN

COMPARISON AND AVERAGING METHODS
IN MATHEMATICAL PHYSICS

ROMA - Istituto Matematico dell'Università

COMPARISON AND AVERAGING METHODS IN MATHEMATICAL PHYSICS
by JAMES SERRIN

It may be worth saying a few words about the general subject of the lectures before beginning with the actual work. I understand that methods of mathematical physics is a subject far too large for any one person to encompass. To the mathematician, on one hand, the subject may mean that part of mathematics which is of immediate or probable value in the study of problems suggested by physics. To the physicist, on the other hand, mathematical physics certainly means the theoretical methods actually used to study the phenomena of mechanics, heat, electromagnetism, field theory, and so forth. These points of view are not entirely separate, of course, but the present lack of communication between mathematician and physicist is sufficient evidence that there is a real difference of emphasis. I think that my point of view in these lectures will contain something of both sides, but will also be rather restricted in its scope. This specialization is essential if one wishes in a few days to come to grips with real problems. The physical side of the lectures will be confined almost completely to continuum mechanics, and even more specially to fluid mechanics, a subject in which I hope there is considerable interest.

The particular topic of comparison methods, which will occupy the first four or five lectures, is itself a large subject. In its most frequent meaning, the phrase "comparison methods" is used in connection with certain problems in the calculus of variations, and with techniques in partial differential equations involving application of the well-known maximum principle. In both cases, the object is to derive <u>inequalities</u> relating quantities of primary physical or mathematical significance. Now to study comparison methods in the calculus of variations would easily require a conference in itself. Therefore, in these lectures we shall concentrate on the comparison method as it appears in connection with the maximum principle in partial differential equations. The particular problems selected for discussion have been

J. Serrin

chosen for their physical interest and in order to exemplify main ideas. Although there are many other problems of importance, it is nevertheless hoped that the techniques illustrated here will be an adequate representation of the field.

I will not comment here on the subject of averaging methods, for these will be discussed in later lectures.

I. THE MAXIMUM PRINCIPLE

This is the generic name for a useful set of theorems in partial differential equations (and hence in mathematical physics) of which perhaps the simplest is the result that a solution of Laplace's equation which assumes an extreme value in the interior of its domain of definition must be a constant Consider more generally elliptic partial differential equations of second order, of the form

$$Lu = a_{ik} u_{ik} + b_i u_i = f(x).$$

Here the coefficients a_{ik}, b_i are <u>bounded</u> functions of $x = (x_1, \ldots, x_n)$ in some domain D of n dimensional Euclidean space, and we have used the abbreviations

$$u = u(x), \qquad u_i = \frac{\partial u}{\partial x_i}, \qquad u_{ik} = \frac{\partial^2 u}{\partial x_i \, \partial x_k},$$

as well as the summation convention. The ellipticity of the operator L is expressed by the condition

$$a_{ik} \xi_i \xi_k \geq m \xi^2 \qquad\qquad (m > 0 $$

for $x \in D$ and all real vectors $\xi = (\xi_1, \ldots, \xi_n)$. Under these conditions, we have the following results, due to Eberhard Hopf.

In three dimensions we shall frequently write x, y, z instead of x_1, x_2, x_3.

J. Serrin

THEOREM (Boundary point lemma). Let S , be an open sphere in D . Let $u = u(x)$ be twice continuously differentiable (class C^2) in S , and continuously differentiable (class C^1) in $S + \{P\}$, where P is a point on ∂S . Assume that

$$\left.\begin{array}{l} Lu \geqslant 0 \\ u < u(P) \end{array}\right\} \quad \text{in } S$$

Then $\partial u / \partial n < 0$ at P , where $\vec{n}$ is any direction into S at P .

The proof is based on a comparison argument, and goes as follows (cf. Courant-Hilbert, Vol. II, pp. 327-38).

Let K be an open sphere internally tangent to S at P (see figure).

Then P is the only maximum point of u in $\overline{K}$, the closure of K . Let the center of K be chosen for the origin of coordinates, and let r_o be the radius of K . Construct a sphere with center P and radius less then r_o , and let C denots the intersection of this sphere with K , (shaded in the figure).

We now introduce the auxiliary comparison function

$$h = e^{-\alpha r^2} - e^{-\alpha r_o^2} \quad ,$$

which is positive in K and vanishes on its boundary. An easy calculation shows that for α sufficient large and $x \in C$

J. Serrin

$$e^{\alpha r^2} Lh = 4\alpha^2 a_{ik} x_i x_k - 2\alpha(a_{ii} + b_i x_i)$$

$$\geq 4\alpha^2 mr^2 - 2\alpha(a_{ii} + b_i x_i) > 0.$$

Now on the lower spherical boundary C_1 of C the function u is <u>bounded away from $u(P)$</u>. Hence there is an $\varepsilon > 0$ such that

$$v = u + \varepsilon h \leq u(P) \qquad \text{on} \quad C_1 .$$

On the upper spherical boundary C_2 of C we have $v = u$. Hence

$$v \leq u(P) \qquad \text{on} \quad C_2 .$$

Consider the function $v = u + \varepsilon h$ in $\overline{C}$. We assert that $v \leq u(P)$. For if not, then v would take on an <u>interior maximum</u> at some point Q C. Then

$$v_i = 0 \text{ at } Q,$$

while

$$(v_{ij}) \text{ is negative definite at } Q .$$

Hence $Lv = a_{ij} v_{ij} \leq 0$ at Q. On the other hand, by hypothesis and by construction $Lv = Lu + \varepsilon Lh > 0$ in C. This contradiction proves the assertion.

Now since $v \leq u(P)$ in C, and $v(P) = u(P)$, it follows that at P

$$\frac{dv}{dn} = \frac{du}{dn} + \varepsilon \frac{dh}{dn} \leq 0 .$$

Since $dh/dn > 0$, we have $du/dn < 0$, and the lemma is proved.

Theorem (Maximum principle). If u satisfies $Lu \geq 0$ in an open connected set R, and has a maximum at an interior point of R, then

J. Serrin

$u \equiv$ constant in R .

Proof. If $u \not\equiv$ constant, and has an interior maximum, it is clear that we can find a sphere S , with closure in R , such that u has a maximum point P on the boundary of S but not interior to S . The hypotheses of the boundary point lemma are satisfied, hence $\partial u/\partial n < 0$ at P , contradicting the fact that the first derivatives of u must vanish at P .

We may restate the result as follows : any non constant function u satisfying $Lu \geqslant 0$ in R takes its maximum only at a boundary point ; and if there is an open sphere in R which is tangent to the boundary at this point, then $\partial u/\partial n < 0$ at this point. (Similar results hold wen $Lu \leqslant 0$, provided we replace "maximum" with "minimum". Finally, if $Lu = 0$, any nonconstant solution u can have neither an interior maximum nor an interior minimum.)

As is well-known, this result allows one to give elementary proof of the uniqueness of the various boundary value problems associated with elliptic partial differential equations (cf. the discussion in Courant-Hilbert). A corresponding comparison theorem is this : Let $T_1(x)$ and $T_2(x)$ be steady state temperature distributions in a bounded domain R . Suppose that $T_1 \geq T_2$ on R . Then $T_1 \geq T_2$ inside R . To see this, we note that $T_1 - T_2$ is a solution of Laplace's equation

$$\Delta u = u_{ii} = 0 \qquad (a_{ik} = \delta_{ik}).$$

Hence $T_1 - T_2$ takes on its maximum and minimum at the boundary, and the result follows. Although this example is elementary, it illustrates the fact that the maximum principle can often lead to results of physical significance by entirely simple means. Let us illustrate this further with examples

J. Serrin

from fluid mechanics.

Consider the Navier-Stokes equations for a voscous incompressible fluid, namely

$$(1) \quad \begin{cases} \operatorname{div} \vec{v} = 0 \\[2mm] \vec{a} = -\operatorname{grad}\left(\dfrac{p}{\rho} + \Omega\right) + \Delta \vec{v}, \qquad \left(\bar{a} = \dfrac{d\bar{v}}{dt},\ \dfrac{d}{dt} = \dfrac{\partial}{\partial t} + \vec{v}\cdot\operatorname{grad}\right). \end{cases}$$

I assume, of course, that you are all acquainted with this equation. The following equations are consequences of (1).

$$(2) \qquad \Delta\left(\frac{p}{\rho} + \Omega\right) = (W^2 - 1)\,\vec{D}{:}\vec{D}$$

$$(3) \qquad \nu\,\Delta H - \vec{v}\cdot\operatorname{grad} H = \nu\omega^2, \quad \text{(steady flow)}.$$

$$(4) \qquad \frac{d\vec{\omega}}{dt} = \vec{\omega}\cdot\operatorname{grad}\vec{v} + \nu\Delta\vec{\omega}$$

Here $\vec{\omega} = \operatorname{curl}\vec{v}$ is the vorticity, $\omega^2 = \vec{\omega}\cdot\vec{\omega}$, $\vec{D}$ is the well known rate of deformation tensor, (The components D_{ij} of D are defined by

$$D_{ij} = \frac{1}{2}\left(\frac{\partial v_i}{\partial x_j} + \frac{\partial v_j}{\partial x_j}\right).\Big)\ \vec{D}:\vec{D} = D_{ij}D_{ij},\ W = |\omega|\big/\sqrt{2D:D}\ \text{is Truesdell's vor-}$$

ticity measure, and $H = \frac{1}{2}q^2 + \frac{p}{\rho} + \Omega$ is the Bernoulli function.

<u>Proof.</u> By vector analysis

$$\vec{a} = \frac{\partial\vec{v}}{\partial t} + \vec{v}\cdot\operatorname{grad}\vec{v} = \frac{\partial\vec{v}}{\partial t} + \vec{\omega}\times\vec{v} + \operatorname{grad}\frac{1}{2}q^2.$$

Hence by an easy calculation, using the fact that $\operatorname{div}\vec{v} = 0$, we have both

$$\operatorname{div}\vec{a} = (1 - W^2)\,\vec{D}:\vec{D},$$

and also

$$\operatorname{div}\vec{a} = \operatorname{div}(\vec{\omega}\times\vec{v}) + \Delta\frac{1}{2}q^2.$$

J. Serrin

But from (1) we have div a = $- \Delta (\frac{p}{\rho} + \Omega)$, and (2) is there fore proved.
Equation (3) requires the identity

$$\text{div} (\vec{\omega} \times \vec{v}) = \omega^2 - \vec{v} \cdot \text{curl } \vec{\omega}$$

$$= \omega^2 + \vec{v} \cdot \Delta \vec{v}$$

$$= \omega^2 + \nu^{-1} \vec{v} \cdot \left\{ \vec{a} + \text{grad} (\frac{p}{\rho} + \Omega) \right\}$$

$$= \omega^2 + \nu^{-1} \vec{v} \cdot \text{grad } H.$$

The last result is the well-known vorticity equation in a viscous fluid (cf. [3]).

Let us now apply the maximum principle to the three equations just derived. The results are as follows :

TH. 1. In a region where the vorticity measure $W \geqslant 1$, the "pressure" $p + \rho \Omega$ cannot have an interior maximum, while in a region where $W \leqslant 1$, $p + \rho \Omega$ cannot have an interior minimum. In case $\Delta \Omega = 0$, as in a uniform gravitational field, $p + \rho \Omega$ can be replaced by p alone.

TH. 2. In steady flow, the Bernoulli function H cannot have an interior maximum.

TH. 3. In steady, plane flow the vorticity

$$\omega = \frac{\partial v}{\partial x} - \frac{\partial u}{\partial y} \qquad\qquad \vec{v} = (u, v, w,)$$

can have neither an interior maximum nor an interior minimum.

Consider next steady irrotational flow of an inviscid incompressible fluid. It is well known that the motion is governed by a velocity potential ϕ

In stating Theorem 1 through 5 we shall for simplicity omit the necessary qualifying phrases concerning the connectivity of the region in question and the non-constancy of the functions involued. The reader may easily supply these things for himself.

J. Serrin

such that $v = \mathrm{grad}\ \phi$ and

$$\Delta \phi = 0$$

By differentiating with respect to x, y, z, in order, we obtain

$$\Delta u = 0, \qquad \Delta v = 0, \qquad \Delta w = 0,$$

where u, v, w, are the velocity components. Hence we have the following results :

TH. 4. The velocity potential ϕ cannot take an interior maximum nor an interior minimum in the region of flow. In steady plane flow, the same holds for the streamfunction , since $= 0$.

TH. 5. The velocity components u, v, w, cannot take either an interior maximum or an interior minimum value in the region of flow. The velocity magnitude q cannot take an interior maximum (since the component of velocity in the direction $\vec{v}$ at the maximum point, could not have a maximum).

Similar results hold when the fluid is compressible, as was shown by Gilbarg. Consider, in particular, steady isentropic, irrotational flow of compressible non-viscous fluid. The velocity potential ϕ satisfies the equation

$$c^2 \Delta \phi - v_i v_j \phi_{ij} = 0 \quad ,$$

where c is the speed of sound, and $\vec{v} = (v_1, v_2, v_3)$. For this equation

$$a_{ij} = c^2 \delta_{ij} - v_i v_j$$

and

$$a_{ij} \xi_i \xi_j = c^2 \xi^2 - (\vec{v} \cdot \vec{\xi})^2 \geqslant (c^2 - q^2) \xi^2 .$$

J. Serrin

Therefore, if the flow is subsonic $(q < c)$ then the equation is elliptic and the maximum principle applies. Since a similar equation holds for the stream function in plane flow, we have thus proved Theorem 4. Now differentiate the velocity potential equation with respect to x. It is not difficult to see (cf. [3] or [6]) that the result is an equation of the form

$$a_{ij} \frac{\partial^2 u}{\partial x_i \, \partial x_j} + \text{linear terms in } \frac{\partial u}{\partial x_i} = 0$$

Thus the velocity components u, v, w, also satisfy the maximum principle, and Theorem 5 is proved.

A somewhat deeper result, which at the same time well illustrates the comparison method, is the following.

TH. 6. Let u be a solution of the Schroedinger type equation

$$Lu = \Delta u + au = 0 \qquad\qquad (n = 3)$$

in some region $r \geqslant r_o$. Suppose that $a = a(x, y, z) \leqslant -m^2$ for some constant $m > 0$, and that

$$u = o\left(\frac{e^{mr}}{r}\right) \qquad\qquad \text{as } r \rightarrow \infty.$$

Then

$$u = 0\left(\frac{e^{-mr}}{r}\right) . \qquad\qquad \text{as } r \rightarrow \infty.$$

The proof goes in essence as follows. The fact that

$$L\left(\frac{e^{-mr}}{r}\right) = (m^2 + a)\,\frac{e^{-mr}}{r} \leqslant 0$$

allows one to construct by standard methods (cf. [4]) a solution v of the differential equation, such that

J. Serrin

$$v = u \qquad\qquad \text{on} \quad r = r_o$$

$$v = 0(e^{-mr}/r) \qquad\qquad \text{as} \quad r \longrightarrow \infty .$$

Now consider the function

$$w = v - u - \varepsilon \frac{e^{mr}}{r} . \qquad\qquad (\varepsilon > 0).$$

We have

$$Lw = - \varepsilon L \left(\frac{e^{mr}}{r}\right) \geqslant 0$$

while also

$$w \leqslant 0 \qquad\qquad \text{on} \quad r = r_o$$

$$w \leqslant 0 \qquad\qquad \text{for } r \text{ sufficiently large.}$$

By the maximum principle, w cannot have an interior positive maximum. Hence $w \leqslant 0$ everywhere. Thus

$$v \leqslant u + \varepsilon \frac{e^{mr}}{r} \qquad\qquad \text{at each point} \quad r \geqslant r_o .$$

Letting $\varepsilon \longrightarrow 0$ yields $v \leqslant u$. By a similar argument $v \geqslant u$. Hence $v = u$ and u has the same asymptotic behavior as v as $r \longrightarrow \infty$. This completes the proof. Other examples will be found in references $[4] - [7]$, as well as in subsequent lectures.

I should like to close this lecture with a statement of the maximum principle as it applies to <u>parabolic</u> partial differential equations. We consider in particular equations of the form

$$\mathscr{L} u = a_{ij} u_{ij} + b_i u_i + c \frac{\partial u}{\partial t} = f(x, t)$$

where a_{ij} and b_i are <u>bounded</u> functions of (x, t) in a domain D of space-

J. Serrin

time, and

$$u = u(x, t), \qquad u_i = \frac{\partial u}{\partial x_i}, \qquad u_{ij} = \frac{\partial^2 u}{\partial x_i \partial x_j}.$$

The fact that the equation is parabolic is expressed by the conditions

$$m \xi^2 < a_{ij} \xi_i \xi_j \qquad , \text{ and } o \leqslant 0.$$

Exactly as in the case of elliptic equations we have the fundamental

Theorem (Boundary point lemma). Let $\tilde{S}$ denote an open sphere in the space of the variables (x, t) , and let S be a set of the form

$$S = \tilde{S} \cap \left\{ t \leqslant t_o \right\}$$

with $S \subset D$. Let P be a point of the spherical boundary of S , not at either t extremity of $\tilde{S}$. Suppose that u is of class C^2 in S , and of class C^1 in $S + \left\{ P \right\}$, and that

$$\left. \begin{array}{c} \mathcal{L} u \geqslant 0 \\ u < u(P) \end{array} \right\} \qquad \text{in } S$$

Then $\partial u / \partial n < 0$, where $\vec{n}$ is any vector directed into S at P .

The proof of this is essentially the same as that of the earlier boundary point lemma, and lemma, and will be omitted.

Theorem (Maximum principle). Let $\tilde{R}$ be an open set in (x, t) space, and let R be a set of the form $\tilde{R} \cap \left\{ t \leqslant t_o \right\}$. Suppose that

$$\mathcal{L} u \geqslant 0 \quad \text{in } R,$$

and that u takes a maximum at $P \in R$. Then $u \equiv$ constant in the set $C(P)$. If $c \leqslant -\mathcal{X}$ where $\mathcal{X}$ is a positive constant, then $u \equiv$ constant in $R(P)$, (see figure).

J. Serrin

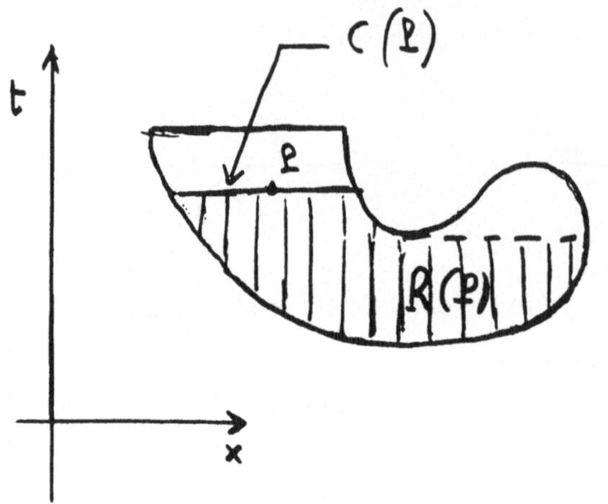

Figure difining the sets C(P) and R(P) ; specifically C(P) denotes that component of the set $\left\{ t = t_p \right\} \cap R$ which contains the point P .

This result is due to Nirenberg [8] . The first statement is a consequence of the boundary point lemma, as in the elliptic case. Indeed supposing that $u \not\equiv u(P)$ on C(P) , there exists some point P' on C(P) such $u(P') < u(P)$, and indeed there is even a vertical segment AP' on which $u < u(P)$. Consider a quarter ellipsoid with semi axes P'A and P'B as shown in the accompanying figure, with B on the same side of P' as P .

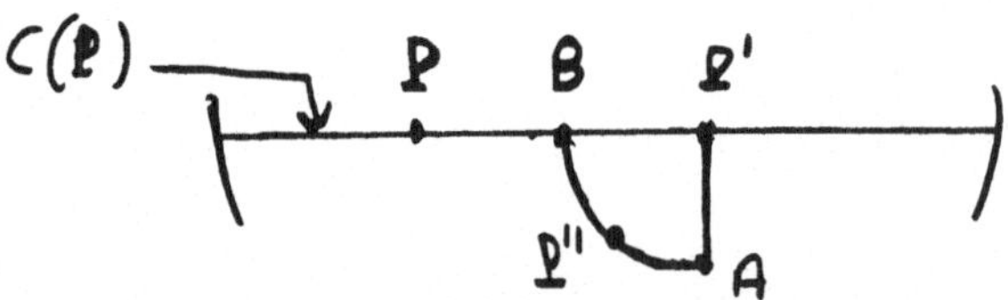

Now as B moves from P' to P there will be a first point where the boundary of the ellipsoid meets a point P'' where $u(P'') = u(P)$. At this point we construct an internal tangent sphere to the ellipsoid. Then by the boundary point lemma $\partial u/\partial n < 0$ at P'' , contradicting the fact that P'' is a spatial maximum of u . (This proof does not take into account all possible configurations of P and P' , but does contain the essence of the argument).

To prove the second part of the theorem, we suppose that $u \not\equiv u(P)$ in R(P) . Then using the first part of the proof it is clear that we can find a point P' such that $u \equiv u(P)$ on C(P') while $u(x, t) < u(P)$ when

J. Serrin

$t_{P'} - \varepsilon \leq t < t_{P'}$. By an argument similar to that of the boundary point lemma, except that the comparison function is of the form

$$h = q(t_{P'} - t) - r^2$$

vanishing on some paraboloid tangent to $C(P')$ at P' , it can be shown that $\partial u/\partial t \geq -\varepsilon \, \partial h/\partial t > 0$ at P' . But on the other hand, since $u \equiv$ constant on $C(P')$, the conditions

$$\mathscr{L}u \geq 0, \qquad\qquad c < 0$$

imply that $\partial u/\partial t \leq 0$ at P' . This contradiction proves the theorem.

Applications of these theorem to the uniqueness of various boundary value problems for parabolic equations are easily given, exactly as in the case of elliptic equations, and we need not dwell on this. In our third and fourth lectures we shall see several less trivial applications. Other applications, in particular to asymptotic behavior, to singular perturbation problems, and to the Stefan problem, are noted in the references, numbers [10] - [14] .

Finally, attention should be drawn to an alternate form of the maximum principle due to Westphal and Nagumo. Its conclusion is somewhat weaker than Nirenberg's but this is compensated for by the greater generality allowed the differential operator $\mathscr{L}u$, which may in particular be strongly non-linear.

References : Chapter I

[1] E. Hopf, Elementare Bemerkungen über die Lösungen Partieller Differentialgleichungen Zweiter Ondnung vom Elliptischen Typus. Sitz. Preuss. Akad. Wiss. 19 (1927), 147-152.

[2] E. Hopf, A remark on linear elliptic equations of second order. Proc. Amer. Math. Soc. 3 (1952), 791-793.

Applications of Comparison Methods to Elliptic Parital Differential Equations

[3] J. Serrin, Mathematical Principles of Classical Fluid Mechanics, Handbuch der Physik, Vol. 8/1. 1959. Esp. §§ 22, 28, 45, 77.

[4] N. Meyers and J. Serrin, The exterior Dirichlet problem for second order elliptic partial differential equations. Journ. Math. Mech. 9 (1960), 513-538.

[5] D. Gilbarg, The Phragmén-Lindelöf theorem for elliptic partial differential equations. J. Rat. Mech. Anal. 1 (1952), 411-417. cf. also J. Serrin, J. Rat. Mech. Anal. 3 (1954), 395-413.

[6] D. Gilbarg, Comparison methods in the theory of Subsonic flows. J. Rat. Mech. Anal. 2 (1953), 233-251.

[7] D. Gilbarg and J. Serrin, On isolates singularities of solutions of second order elliptic differential equations. J d'Anal. Math. 4 (1956), 309-340.

J. Serrin

Maximum Principles for Parabolic Partial Differential Equations

[8] L. Nirenberg, A strong maximum principle for parabolic equations. Comm. Pure Appl. Math. $\underline{6}$ (1953), 167-177. cf. also A. Friedman, Pacific J. Math. 8 (1958), 201-211.

[9] H. Westphal, Zur Abschatzung der Losung nichtlinearer parabolischer Differentialgleichungen. Math. Z. $\underline{51}$ (1949), 690-695.

Applications of Comparison Methods to Problems
Involving Parabolic Partial Differential Equations

[10] M. Krzyzanski, Sur les solutions de l'equation linéare du type parabolique déterminées par les conditions initiales. Ann. Soc. Polon. Math. $\underline{18}$ (1945), 145-156.

[11] D. Aronson, Linear Parabolic differential equations containing a small parameter. J. Rat. Mech. Anal. $\underline{5}$ (1956), 1003-1014.

[12] R. Narasimhan, On the asymptotic stability of solutions of parabolic differential equations. J. Rat. Mech. Anal. $\underline{3}$ (1954), 303-314.

[13] A. Friedman, Free boundary problems for parabolic equations I. II, III. Journ. Math. Mech. $\underline{8}$ (1959), 499-517; 9 (1960), 19-66, 327-346.

[14] R. Redheffer, Bemerkungen über Monotonie und Fehlerabschätzung bei nichtlinearen partiellen Differentialgleichungen. Arch. Rat. Mech. Anal. $\underline{10}$ (1962), 427-457.

Maximum Principles and Comparison Methods for Other Types of Equations

[15] S. Agmon, L. Nirenberg, and M. Protter. Comm. Pure Appl. Math. $\underline{6}$ (1953), 445.

J. Serrin

[16] C. Morawetz, Note on a maximum principle and a uniqueness theorem for an elliptic hyperbolic equation. Proc. Roy. Soc. London A 236 (1956), 141-144

[17] C. Pucci, Proprietà di massimo e minimo delle soluzioni di equazioni a derivate parziali del secondo ordine di tipo ellittico e parabolico. Rend. Lincei. Ser. 6, 23, 24.

In this as in later chapters, the reference list is mainly restricted to papers actually quoted. Further references may be found by consulting the bibliographies of the papers quoted, as well as the second volume of "Method of Mathematical Physics", by Courant and Hilbert.

J. Serrin

II HYDRODYNAMICAL COMPARISON THEOREMS

In this lecture we shall be concerned almost exclusively with steady, plane, irrotational flows of an incompressible fluid. As we have already remarked, such flows are characterised by a stream function ψ such that the velocity components are given by

$$u = \psi_y , \qquad\qquad v = -\psi_x ,$$

and

$$\psi_{xx} + \psi_{yy} = 0 .$$

Moreover, by its very definition, the streamlines of the motion are the curves ψ = constant.

Our main interest will be in the well konwn Helmholtz-Kirchhoff free streamline theory, and most particularly in the case where a constant pressure wake or cavity forms behind an obstacle in uniform flow, (the speed on the bounding streamline is then constant by Bernoulli's theorem).

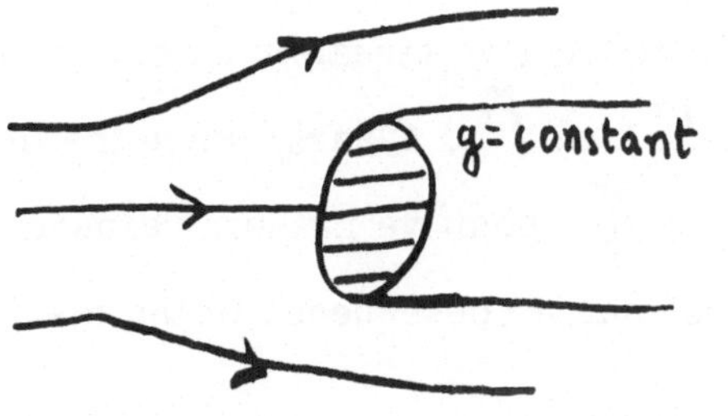

I assume that you are familiar with the fundamental theory of this flow model, as found for example in the books of Lamb and Milne-Thomson, including the hodograph method for the determination of explicit solutions. Here we shall be concerned with the associated comparison theory, which yields a variety of interesting qualitative results and adds greatly to the understanding of the phenomena involved. It should be emphasized also that the comparison theory applies to more or less general obstacle shapes, and is not confined to the more special flows where an explicit solution is available. Unfortunately, a

J. Serrin

similar body of results has not yet been developed for other classes of free boundary problems, say those involving free surfaces under gravity.

The teory is ultimately based on two fudamental hydrodynamical comparison theorems, which we now present.

THEOREM (Speed comparison) Let R and $\bar{R}$ be flow regions for two plane flows having uniform non-zero velocities q_o and q_o at infinity, where $\bar{q}_o \geqslant q_o$. Let R and $\bar{R}$ be bounded by smooth streamlines extending to $x = \pm \infty$, as shown. If $R \subset \bar{R}$ and the boundaries touch at P, then

$$\bar{q}(P) \geqslant q(P),$$

the equality holding only if $R = \bar{R}$ and $\bar{q}_o = q_o$.

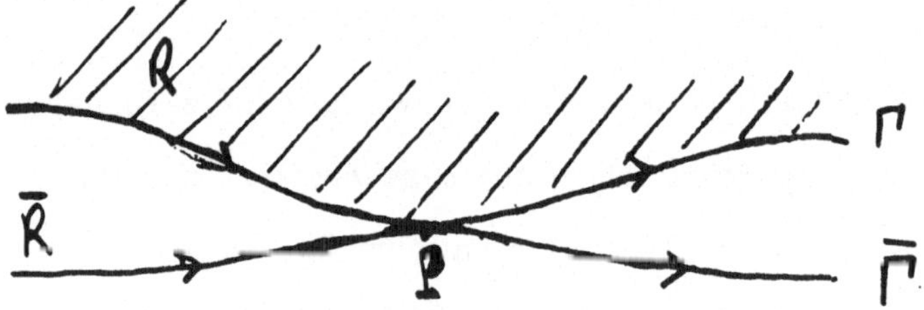

Proof. Let ψ and $\bar{\psi}$ be the respective streamfunctions for the flows in question, assumed to be zero on Γ and $\bar{\Gamma}$. Clearly since the flow is from left to right, we have both ψ and $\bar{\psi}$ positive in their respective flow regions (technically this can be obtained as a consequence of the maximum principle for the streamfunction).

Consider first the case when $\bar{q}_o > q_o$. Since

$$\frac{\partial (\bar{\psi} - \psi)}{\partial y} = \bar{u} - u \to \bar{q}_o - q > 0 \qquad \text{as } (x, y) \to \infty,$$

it is clear that $\bar{\psi} - \psi > 0$ outside some large circle. But also $\bar{\psi} - \psi \geqslant 0$ on Γ. Therefore, again by the maximum principle (since $\Delta(\bar{\psi} - \psi) = 0$),

J. Serrin

we find

$$\bar{\psi} - \psi > 0 \qquad \text{in } R.$$

By the boundary point lemma (since $\bar{\psi} - \psi = 0$ at P) we have

$$\bar{q}(P) - q(P) = \frac{\partial}{\partial n} (\bar{\psi} - \psi) > 0.$$

Next suppose $\bar{q}_0 = q_0$. If $R = \bar{R}$ then $\bar{q}(P) = q(P)$, and we are done. Suppose then that $R \ne \bar{R}$. We observe that the stream function

$$\alpha \cdot \bar{\psi} , \qquad (\alpha > 1)$$

describes a flow in $\bar{R}$ with velocity at infinity $\alpha q_0 > q_0$. Hence by the previous argument

$$\alpha \bar{\psi} - \psi > 0.$$

Letting $\alpha \to 1$ yields $\bar{\psi} - \psi \geqslant 0$. But $\bar{\psi} - \psi$ is not constant, hence it cannot take its minimum in R . That is

$$\bar{\psi} - \psi > 0 \qquad \text{in } R.$$

The conclusion $\bar{q}(P) > q(P)$ now follows as before, and the theorem is proved. For references, see $[3]$, $[5]$, and $[6]$.

THEOREM (Interchange theorem). Let two non-zero plane flows be defined in regions R and $\bar{R}$ bounded by smooth streamlines Γ and $\bar{\Gamma}$ extending to infinity. Suppose Γ and $\bar{\Gamma}$ have an arc MN in common, but interchange their positions on either side of MN , as shown. Then

$$\frac{\bar{q}(M)}{\bar{q}(N)} \gtrless \frac{q(M)}{q(N)} ,$$

the equality holding only if $R = \bar{R}$.

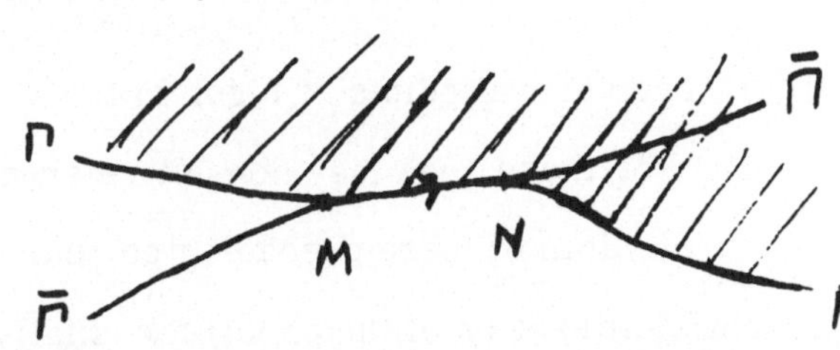

J. Serrin

     Proof. If $R = \bar{R}$ the two flows must be multiples of one another, and the equality is obvious. Suppose then that $R \nmid \bar{R}$, and assume also that $\bar{q}(N) = q(N)$, (this can always be attained by multiplication of one of the flows by a suitable factor, a process which leaves the conclusion invariant.) It is thus necessary to prove under these circumstances that

$$\bar{q}\,(M) > q(M)\,.$$

Consider the function $\Omega = \bar{\psi} - \psi$. Since

$$\frac{\partial \Omega}{\partial n}(N) = \bar{q}(N) - q(N) = 0\,,$$

it is clear that a level line $\Omega = 0$ of Ω issues from N into $R \cap \bar{R}$. Some rather annoying difficulties are avoided if we assume that this level line C extends directly to ∞ without intersecting the boundary of $R \cap \bar{R}$ at any time. It is then the case that the set $R \cap \bar{R}$ is divided into two portions A_1 and A_2 such that $\Omega \geqslant 0$ on the finite boundary of A_1 and $\Omega \leqslant 0$ on the finite boundary of A_2 . An application of the maximum principle in the separate regions A_1 and A_2 yields then $\Omega > 0$ in A_1 and $\Omega < 0$ in A_2 . In particular $\Omega > 0$ near M , hence by the boundary point lemma

$$\bar{q}(M) - q(M) = \frac{\partial}{\partial n}\,(\bar{\psi} - \psi) > 0.$$

Q.E.D.

There are several points of rigor in the above proof which require additional effort. For these, one may consult reference $\begin{bmatrix}6\end{bmatrix}$ or $\begin{bmatrix}8\end{bmatrix}$.

     Both of these preceding comparison theorems remain true (without essential alteration of the proof) for axially symmetric flows, and for subso-

J. Serrin

<u>nic</u> flows of a compressible fluid. The reason is that the stream function ψ satisfies an elliptic equation in either of these more general situations, while the proofs were based only on comparison arguments involving ψ . Of course, the proofs above require the maximum principle only in its simple form for solutions of Laplace's equation, while in the more general situations it is necessary to have the maximum principle in its general form for elliptic equations. The application of comparison methods in subsonic flow is due particularly to Gilbarg.

<u>Applications of the speed comparison theorems</u>. As a first extremely simple observation consider irrotational flow past a symmetric obstacle, as shown in the figure.

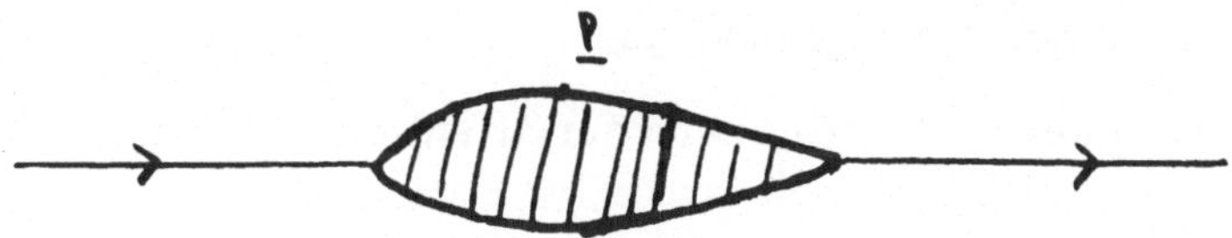

Then certainly the maximum flow velocity is greater than the speed at infinity (apply the speed comparison theorem at P). Suppose next that the forward part of the obstacle is in the form of a wedge:

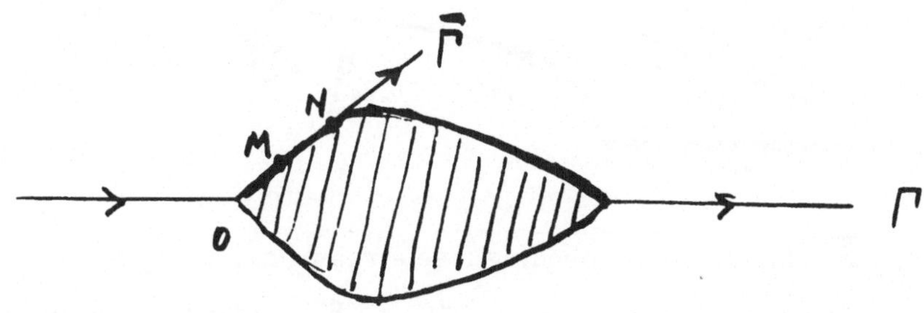

Comparing the given flow with that in the wedge bounded by $\bar{\Gamma}$, and using

J. Serrin

the interchange theorem, we have

$$\frac{\bar{q}(M)}{\bar{q}(N)} > \frac{q(M)}{q(N)} \quad .$$

Letting s be the arc length from 0 , setting $q(N) = q(s_0)$, $q(M) = q(s)$, and observing that then $\bar{q}(M) / \bar{q}(N) = \text{const. } s^{\frac{\alpha}{\pi-\alpha}}$, we get

$$q(s) < q(s_0)\left(\frac{s}{s_0}\right)^{\frac{\alpha}{\pi-\alpha}} \quad , \qquad s \lessgtr s_0 \ .$$

a result which is not directly obvious.

From here on, let us turn our attention to the main issue of free boundary flows.

Consider a symmetric infinite cavity flow (the upper symmetric part is sufficient), and let T denote the curve consisting of the upstream axis of aymmetry together with the obstacle up to the detachment point A .

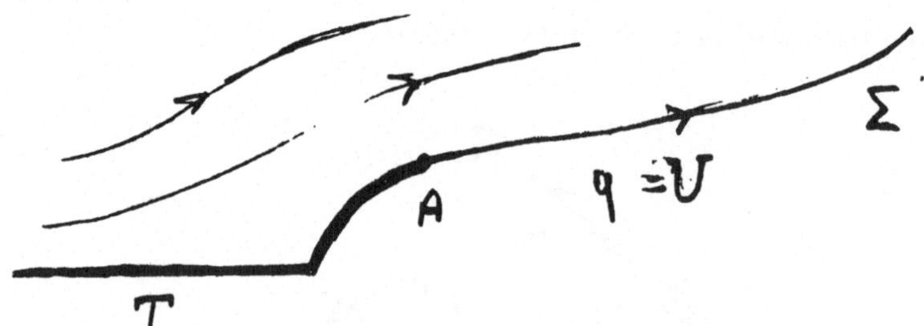

Also let Σ be the corresponding free streamline, along which the velocity is assumed to have a constant value. Since Σ extends down stream to infinity, it is apparent that this value is precisely the stream speed, which we denote here by U . The corresponding flow is of the type to which the preceding speed comparison theorems apply. This will in fact the basis for the following results.

Single intersection theorem. Any straight line which does not cut T can intersect Σ in at most one point [7] .

Proof. Suppose the contrary, for example as in the following figure.

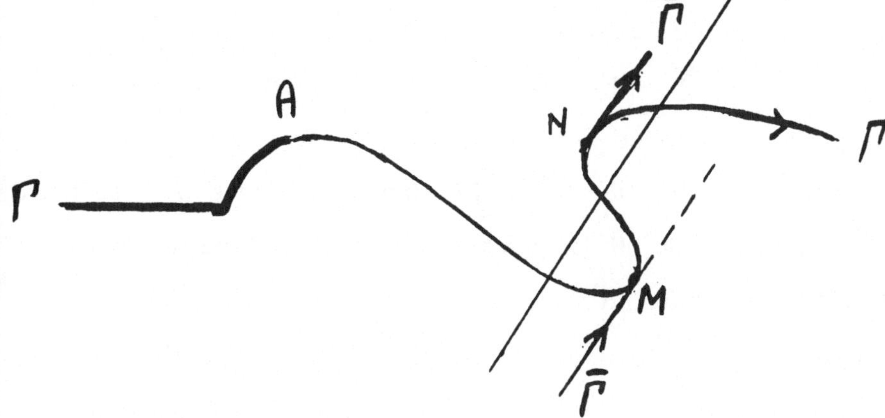

By applying the interchange theorem to this situation one sees that

$$\frac{\bar{q}(M)}{\bar{q}(N)} > \frac{q(M)}{q(N)} = 1.$$

However, by the speed comparison theorem it is evident that

$$\bar{q}(M) < U \qquad \text{and} \qquad U < \bar{q}(N).$$

This yields $\bar{q}(M) / \bar{q}(N) < 1$, in contradiction with the preceding inequality. Since other cases of multiple intersection can obviously be handled by a similar technique, the theorem is proved.

Uniqueness theorem. Consider symmetric flows past an obstacle C (in the upper half plane) with the free streamline detaching from a fixed endpoint A on C . Suppose that the corresponding curve T is star-

J.Serrin

like with respect to some point Q on or below the axis of symmetry. Then if Σ and $\bar{\Sigma}$ are two corresponding free streamlines, we have $\Sigma = \bar{\Sigma}$ and the flows are identical [5], [6], [7].

Proof. The single intersection theorem shows that $T + \Sigma = \Gamma$ and $T + \bar{\Sigma} = \bar{\Gamma}$ are both starlike with respect to Q. In the standard theory of free boundary problems it is shown that the free boundaries Σ and $\bar{\Sigma}$ are asymptotic to a parabola downstream. Now suppose $\Sigma \neq \bar{\Sigma}$. Then one of the curves, say Σ, lies above the other at infinity, or else they are both asymptotic to the same parabola. In either case, a similarity transformation (contraction) about Q takes $\bar{\Gamma}$ into a new curve $\bar{\Gamma}'$ such that $\bar{\Gamma}'$ and Γ have a point P in common, and such that the flow region $\bar{R}'$ bounded by $\bar{\Gamma}'$ contains R.

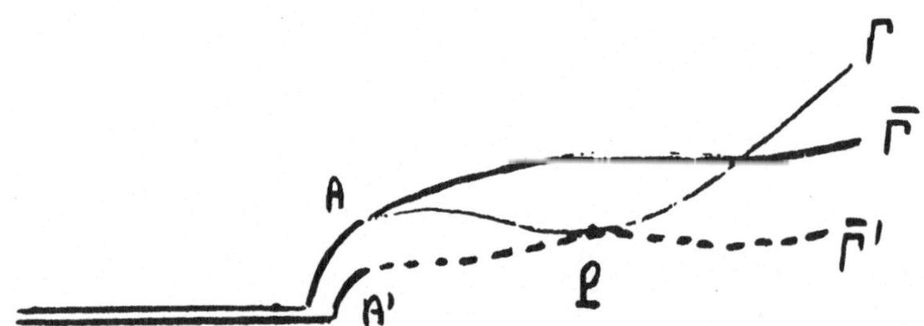

Moreover the flow in $\bar{R}'$ has the same velocity at corresponding points as the original flow in $\bar{R}$. The speed comparison theorem applied to these two fows thus yields

$$U = \bar{q}(P) > q'(P) = U.$$

This contradiction shows that we must actually have $\Sigma = \bar{\Sigma}$, and the theorem is proved.

Interchange Theorem. Let $\bar{C}_1$ and C_2 be two obstacles for the

J.Serrin

symmetric infinite cavity problem, with the same detachment point   A   for each flow. Suppose that $\bar{T}$ (or T) is starlike with respect to a point Q , and that both   T   and $\bar{T}$ lie to the left of the extended line AQ.   Then if T   lies above $\bar{T}$, we have Σ below $\bar{\Sigma}$, [5] , [7] .

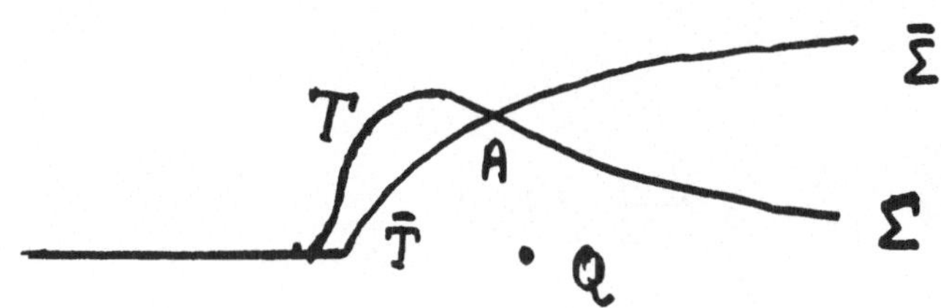

Proof. We know that $\bar{\Sigma}$ and Σ are asymptotic to parabolas. It will first be shown that Σ lies below $\bar{\Sigma}$ at infinity. For otherwise (if Σ lies above $\bar{\Sigma}$ or if they are both asymptotic to the same parabola) then $\bar{R}$ can be contracted so that $\bar{R}'$ contains R , and their boundaries touch at a point P . The speed comparison theorem, applied exactly as in the preceding proof, supplies a contradiction.

Having shown this, it follows that if the theorem were not true the two flows would be as shown in the figure.

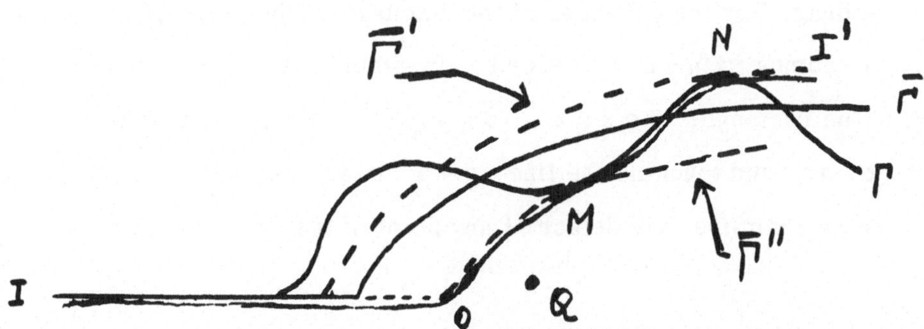

We construct three auxiliary flow regions. First by an expansion we construct $\bar{R}'$. Next by the contraction we get $\bar{R}''$. Finally $\bar{R}'''$ is the region bounded by IOMNI' .   .Applying the speed comparison theorem at   M   one obtains

J. Serrin

$$\overline{q'''}(N) < \overline{q''}(M) = U,$$

and similarly

$$\overline{q'''}(N) > \overline{q'}(N) = U .$$

Next applying the interchange comparison theorem to R   and $\overline{R'''}$ we have

ve

$$\frac{\overline{q'''}(M)}{\overline{q'''}(N)} \geqslant \frac{q(M)}{q(N)} = 1.$$

The last three conditions are in contradiction, and the theorem is proved. The proof is entirely the same if the curve   T   is starlike.

Similar results hold for axially symmetric flows and for subsonic flows of a compressible fluid ([7] - [10]), and for free boundary problems involving jets. The proofs in these cases, although similar in their basic structure, tend to be technically more difficult and for this reason have not been given in detail.

The preceding theorems allow an interesting application to the problems of determining the symmetric obstacle of given dimensions with least cavity drag. For the purposes of the discussion, the class of obstacles allowed into competition will be described by smooth starlike curves   C , confined to the rectangle $0 \leqslant x \leqslant a$, $0 \leqslant y \leqslant b$, joining the origin to a point on $x = a$, and touching the line $y = b$ in at least one point. The associated free streamline may detach at any point of the obstacle   C   provided that

i) it does not reintersect   C

ii) the max flow speed is   U .

(This condition reflects the assumption that cavitation will immediately occur at pressures lower than that in the free stream.)

J. Serrin

We consider a particular curve K in this class constructed by choosing the Kirchhoff infinity cavity flow past a vertical segment in such a way that the free streamline passes through the point (a, b) ; see figure.

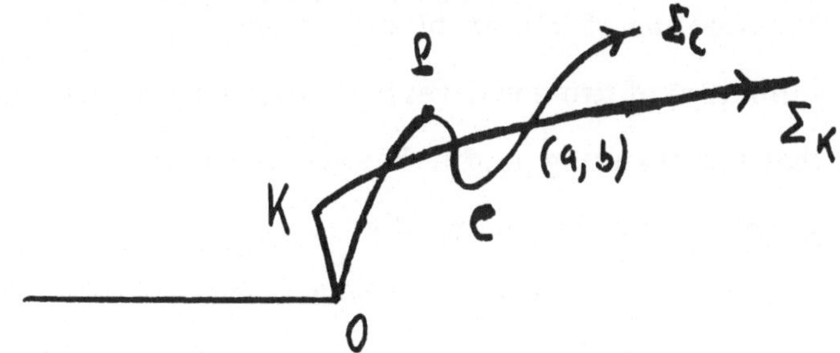

We assert that the curve K has less cavity drag then any other curve C in the class described above. $[5]$, $[7]$.

    <u>Proof.</u> If C lies below K , then by the interchange theorem, Σ_C is above Σ_K . But it is well known that the drag increases as the cavity opening increases (see $[2]$), whence Drag $(C) >$ Drag (K).

    If C lies partially above K , then there is a vertical segment L whose free streamline Σ is tangent to C , say at P . Clearly Σ_C must cross Σ for otherwise the speed comparison theorem gives a contradiction to ii) at P . But if Σ_C crosses Σ it then stays above Σ from then on, because of the interchange theorem. It follows then, as before, that Drag $(C) >$ Drag $(K_L) >$ Drag(K).

    The result obviously applies similarly to axisymmetric flow. Thus the design of an obstacle of least cavity drag involves, rather remarkably, a flat leading profile.

    As a final application we notice a remarkable relation between free boundary problems and the problem of determining the symmetric obstacle of given dimensions for which the <u>maximum flow speed</u> is <u>least</u>. To state the problem quite definitely, consider smooth, symmetric obstacles with a fixed

J. Serrin

ratio of width to length, placed in a uniform stream with velocity U   at infinity. Among such obstacles, the problem is to determine one for which the maximum flow velocity (necessarily attained on the profile) is least.

We assert that the solution of this problem is a profile E   consisting (above the axis of symmetry) of two equal vertical segments joined by a convex arc S , such that the resulting profile has the prescribed dimensions and has the property that the corresponding flow is of <u>constant speed</u> on S . (Thus S   is the solution of a certain free boundary problem. It is easy to show by the hodograph method that there exists exactly one solution of this problem; indecd, the problem and solution are identical with that of the celebrated Riabouchinsky finite cavity flow. Curiously, although this free boundary model is physically unrealistic in its original setting, as a solution of a cavitation problem, here it proves to have a genuine physical importance.) To prove that the profile E   is the required solution of the given problem, consider any obstacle B $\neq$ E   with the same width to length ratio. We may suppose that B   has the same length as E   by an appropriate normalization, (leaving the speeds at corresponding points unchanged.) Then since B   has the same width as E   it must either touch or cross the arc S   at some point. If it touches S , say at P , then by the speed comparison theorem

$$q_B(P) > q_E(P) = \text{max speed in flow past}\ \ E\ .$$

Thus the maximum speed in the flow past B   is greater than that of the flow past E .

J. Serrin

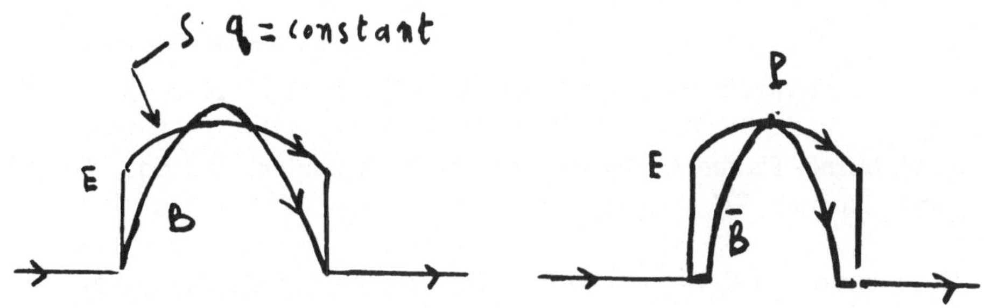

On the other hand if B crosses E , then by a contraction of B into a smaller set $\bar{B}$ we can manage that $\bar{B} \subset E$ while $\bar{B}$ touches S at a point P . The speed comparison theorem then applies exactly as before, and the assertion is proved. This result was first obtained by Gilbarg and Shiffman, in a slightly different physical situation. It is also easy to show (see [4] that as the length-width ratio is <u>increased</u> the speed on S must <u>decrease</u> . Thus the longer the obstacle is allowed to be, for a given width, the more one can reduce the maximum flow speed (toward its limiting value U).

J.Serrin

REFERENCES :   CHAPTER 2
Texts and Monographs

[1] H.Lamb, Hydrodynamics, 6th edition, Cambridge, 1932, pp. 94-105.

[2] L.M.Milne-Thomson, Theoretical Hydrodynamics, 3rd Ed., New York, 1950; Chapter XII.

[3] G.Birkhoff and E.Zarantonello, Jets, Wakes and Cavities, New York, 1957.

[4] D.Gilbarg, Jets and cavities, Handbuch der Physik, Vol.9. Springer, 1960. Esp. 368-387, 407-409.

Papers

[5] M.Lavrentieff, Sur certaines propriétés des fonctions univalentes et leur applications à la théorie des sillages. Mat. Sbornik 46 (1938), 391-458. (Russian, with French summary).

[6] J.Serrin, Uniqueness theorems for two free boundary problems. Amer. J.Math. 74 (1952), 492-506.

[7] J.Serrin, On plane and axially symmetric free boundary problems. J. Rat. Mech. Anal. 2 (1953), 563-575.

[8] J.Serrin, Comparison theorems for subsonic flows, J.Math. Physics, 33 (1954), 27-45.

[9] D.Gilbarg, Uniqueness of axially symmetric flows with free boundaries. J.Rat. Mech. Anal. 1 (1952), 309-320.

[10] D.Gilbarg, Comparison methods in the theory of subsonic flows. J.Rat. Mech. Anal. 2 (1953), 233-251. Cf. also M.Schiffer, Analytical theory of subsonic and supersonic flows. Handbuch der Physik, Vol.9. Springer 1960; Especially pp. 106-107.

[11] D.Gilbarg and M.Shiffman, On bodies achieving extreme values of the critical Mach number. J.Rat. Mech. Anal. 3 (1954), 209-230.

J. Serrin

III. THE COMPARISON METHOD IN BOUNDARY LAYER THEORY

A. More than 60 years have passed since Prandtl formulated the main concepts of boundary layer theory in 1904. But today we are still far from having a complete knowledge of those equations. Even the simplest case of a steady laminar two dimensional boundary layer in an incompressible fluid presents many mathematical problems, for example, existence, stability, and asymptotic behavior, which are still only partially solved.

At the same time, among many mathematicians there is the feeling that boundary layer theory, being based on an approximation, is in some way not quite respectable. Perhaps this is partly due to the lack of established techniques for treating the boundary layer equations as a mathematical system, or in other words, to the fact that the equations themselves are even further approximated in most of the known solution processes. However, in 1958 Karl Nickel, using comparison methods in the theory of parabolic differential equations, was able to obtain a number of valuable results concerning the exact theory. These results I wish to discuss in this and the following lecture, though the presentation itself owes much to a paper of Velte.

I begin by reminding you that classical boundary layer theory is based on the assumption that in viscous flow past a fixed wall, the flow region can be conveniently divided into two parts:

1) A thin layer adjacent to the body, in which the fluid velocity rapidly adjusts from zero to its general streaming value. In this region the gradients of the tangential velocity component will be large and the viscous shearing stresses will exert appreciable influence on the motion.

2) The remaining region of flow in which the effects of viscosity are of much less immediate importance. In this region the fluid may be con-

J. Serrin

sidered inviscid, and potential theory is usually applied to determine the flow. It should be observed that this flow in the exterior region automatically provides the streaming velocity required in 1).

By these assumptions, the flow problem is thus divided into two different and separate parts. We observe that an approximation process is explicitly postulated for the second region, the full Navier-Stokes equations

$$\operatorname{div} \vec{v} = 0$$

$$\frac{d\vec{v}}{dt} = -\operatorname{grad}\left(\frac{p}{\varrho} + \Omega\right) + \nu\, \Delta\, \vec{v}$$

being <u>replaced</u> by their inviscid counterparts

$$\operatorname{div} \vec{v} = 0, \qquad \operatorname{curl} \vec{v} = 0 \qquad\qquad (\text{i.e. } \Delta\, \phi = 0).$$

In the thin boundary layer region a different approximation is necessary, which, while greatly simplifying the Navier-Stokes equations, still retains the effects of viscosity. This approximation lies at the heart of boundary layer theory, and consists in neglecting certain terms in the Navier-Stokes equations. If coordinates are introduced as shown (I confine myself from

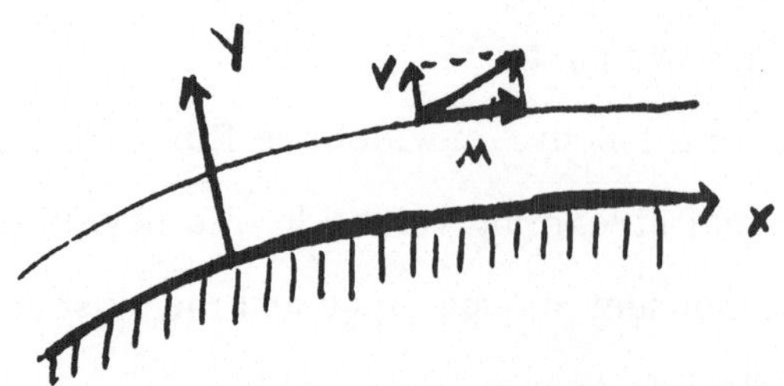

here on to steady plane flow, though the ideas in many cases carry over to the unsteady three dimensional case), the approximation is given by

J. Serrin

$$u_x + v_y = 0,$$

$$uu_x + vu_y = -\frac{1}{\rho}\frac{dp}{dx} + \nu\, u_{yy} \;,$$

where $p = p(x)$ is the pressure (1) in the streaming flow, determined according to Bernoulli's equation

$$-\frac{1}{\rho}\frac{dp}{dx} = UU_x \;, \qquad U(x) = \text{streaming velocity.}$$

We thus have the familiar Prandtl system of equations

$$uu_x + vu_y = UU_x + \nu\, u_{yy}$$

$$u_x + v_y = 0,$$

which I assume that you have all seen at one time or another. The associated problem of partial differential equations is that of the <u>downstream evaluation of the u-velocity profile</u>. We thus suppose that at some initial position, say $x=0$, we have

$$u(0, y) = \tilde{u}(y) \;\; \text{given,}$$

together with the well-known Prandtl boundary conditions

$$u = v = 0 \qquad \text{when}\;\; y = 0$$

$$u = U \qquad \text{when}\;\; y = \infty \qquad \text{(at edge of the boundary layer).}$$

---

(1) More precisely, p here corresponds to $p + \rho\Omega$ previously.

J. Serrin

In part B I shall discuss this latter condition further.

The problem as thus phrased bears a striking resemblance to the theory of parabolic differential equations, with the coordinate x playing the role of the time t . That is, the <u>downstream evolution</u> of the velocity profile is analogous to a <u>time evolution</u> problem.

In the following results we shall apply the maximum principle (cf. lecture 1) to the study of this problem. I shall assume throughout that. $u \geqslant 0$ for any solution in question, and also that the streaming velocity U is of class C^1 and positive for $x \geqslant 0$. Finally, we shall use the letter R to denote a region in the boundary layer of the form

$$R = \left\{ 0 < x \leqslant x_0 \ , \qquad 0 < y < \infty \right\} .$$

THEOREM 1. Suppose $U_x \geqslant 0$. Then $u > 0$ in R . Moreover, $u_y(x, 0) > 0$ for $0 < x \leqslant x_0$.

<u>Proof</u> . For a <u>given solution</u> $u = u(x, y)$, $v = v(x, y)$ for the boundary layer equations, let $\mathcal{L}$ denote the <u>linear differential operator</u>

$$\mathcal{L} \equiv \frac{\partial^2}{\partial y^2} - u \frac{\partial}{\partial x} - v \frac{\partial}{\partial y} , \qquad (c = -u \leqslant 0).$$

Then one sees at once that

$$\mathcal{L} u = -U U_x \leqslant 0.$$

We can thus apply the maximum principle in the form stated in the opening lecture. That is, suppose for contradiction that $u = 0$ at some point P of R . Then u has a minimum at P , and consequently u is constant on

J. Serrin

C(P) . That is

$$u \;=\; 0 \qquad \text{on} \quad C(P)$$

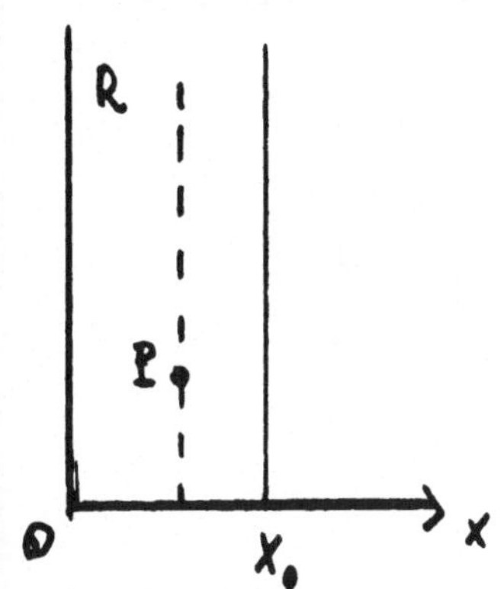

This violates the condition at infinity, and the first part of the theorem is proved. The second part of the theorem now follows from the boundary point lemma in an obvious way.

This result shows that in a flow with $u \geqslant 0$ and $U_x \geqslant 0$, no incipient backflow can develop, and the separation condition $u_y(x, 0) = 0$ can never arise. (This theorem is in fact the theoretical justification for the term favorable pressure gradient.)

THEOREM 2. The shear component $\mu\, u_y$ cannot assume an extremum at any point of R .

Proof. This is based on the calculation

$$0 = \frac{\partial}{\partial y}\left(\nu\, u_{yy} + UU_x - uu_x - vu_y\right)$$

$$= \nu\, u_{yyy} - u_y u_x - uu_{xy} - v_y u_y - vu_{yy}$$

$$= \nu\, u_{yyy} - uu_{yx} - vu_{yy} = \mathcal{L}\, u_y \;.$$

Thus if u_y should assume an extremum at $P \in R$, then $u_y = $ constant on C(P) . Integration then yields

$$u \;=\; ay + b \qquad \text{on} \quad C(P),$$

J. Serrin

which is in contradiction to the assumed boundary conditions.

THEOREM 3. Suppose that $\lim_{y \to \infty} u(x, y) = U(x)$ uniformly in x . Suppose also that the initial velocity profile $\tilde{u}(y)$ satisfies the condition

$$\tilde{u}(y) \leqslant U(0) + e$$

where $e \geqslant 0$ is the initial <u>overvelocity</u> . Then for $(x, y) \in R$

$$u(x, y) < \sqrt{U(x)^2 + 2eU(0) + e^2}.$$

In paticular, if there is no initial overvelocity, then $u < U$, and no over-velocity will developat a later time.

<u>Proof</u>. Set $Z = U^2 - u^2$, in order to <u>compare</u> U and u . Then

$$\mathscr{L} Z = \nu(U^2 - u^2)_{yy} - u(U^2 - u^2)_x - v(U^2 - u^2)_y$$

$$= \nu(-2uu_y)_y - 2u(UU_x - uu_x) - 2uvu_y$$

$$= -2u(\nu u_{yy} + UU_x - uu_x - vu_y) - 2\nu u_y^2 = -2\nu u_y^2 .$$

Thus

$$\mathscr{L} Z \leqslant 0,$$

and the maximum principle may be applied to Z ;that is, if Z takes a minimum at a point P of R , then $Z \neq$ const. on $C(P)$.

Now we observe that

J. Serrin

$$\begin{cases} Z = U(0)^2 - \tilde{u}(y)^2 \geq -2eU(0) - e^2 & \text{on} \quad x = 0 \\[2mm] Z = U^2 > 0 & \text{on} \quad y = 0 \\[2mm] \lim Z = 0 & \text{as} \quad y \to \infty, \text{ uniformly in x.} \end{cases}$$

Since it is impossible for Z to be constant on any line $C(P)$, we infer that Z cannot have a minimum in R . Consequently, the boundary conditions require that

$$Z > -2eU(0) - e^2 \quad .$$

Q.E.D.

We remark that a similar result holds in three dimensional boundary layer theory. Consider in particular flow over a plate in the x, y plane, the coordinate z being taken in the direction normal to the plate. Then if $u^2 + v^2 \leq U^2 + V^2$ on some arc AB of the initial curve, as shown, we have $u^2 + v^2 \leq U^2 + V^2$ in the corresponding shaded region.

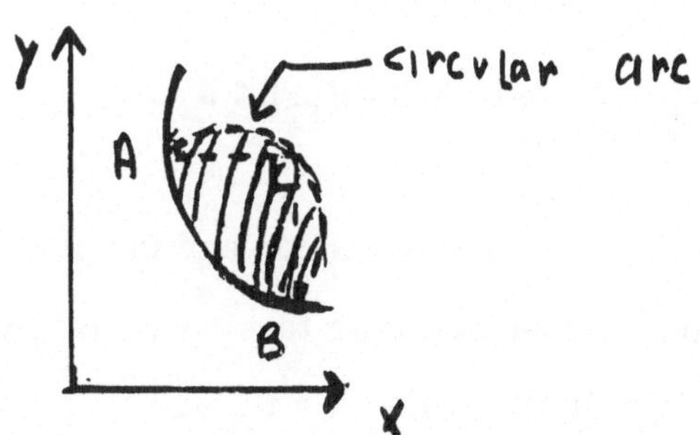

The next theorem shows another interesting property of the boundary layer equations.

THEOREM 4 . Suppose that $\lim\limits_{y \to \infty} u_y = 0$ uniformly in x , and that the initial profile $u(y)$ has exactly N proper extrema. Then any down-

---

A proper extremum is a point where a function takes on a strict local maximum or a strict local minimum.

J. Serrin

stream profile can have at most N proper extrema. If $\tilde{u}_y \geqslant 0$ then $u, u_y > 0$ in R.

Proof of Theorem. The last statement follows from Theorem 3 and the condition $\lim\limits_{y \to \infty} u_y = 0$, (since $u_y \geqslant 0$ on $y = 0$). To prove the first part suppose that the graph of u at $x = x_0$ had more than N extrema. Then we can find $N + 1$ points P_i on the line $x = x_0$ such that, with respect to the variable y,

$$u \text{ has a max at } P_1, P_3, \ldots\ldots$$
$$u \text{ has a min at } P_2, P_4, \ldots\ldots$$

Between P_1 and P_2 on $x = x_0$ there is a point P_{12} where u_y takes a negative minimum with respect to its values on the segment P_1, P_2. But since u_y cannot take a minimum on R, it is clear that a level line C_1

$$u_y = \text{Const.} = u_y(P_{12}) < 0$$

issues from P_{12} into R. This line cannot end in R, since u_y has a maximum principle, and neither can it go to $y = 0$ or $y = \infty$. It therefore finds its way back to the initial line $x = 0$.

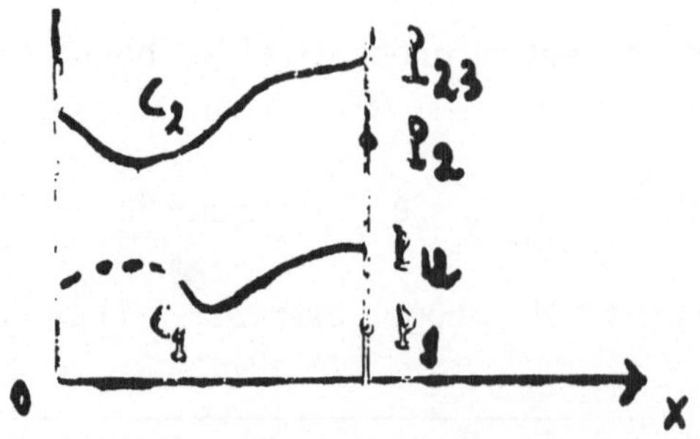

J.Serrin

Apllying the same process between P_2 and P_3 we see that from some point P_{23} three issues a level line C_2

$$u_y = \text{Const.} = u_y(P_{23}) > 0$$

which likewise must find its way back to $x = 0$. Continuing in this way we obtain N level curves each lying above the proceding, and each extending back to the initial line.

We construct lnslty the curve C_{N+1} . Since $u_y(P_{N+1}) = u_y(\infty) = 0$, there will be some point $P_{N+1,\,\infty}$ on $x = x_o$ at which u_y assumes <u>with respect to $x = x_o$</u> an extremum opposite in sign from that at $P_{N,\,N+1}$. We can then construct the level curve C_{N+1} exactly as before.

We now have $N + 1$ points on the initial profile at which u_y is alternately negative, positive, negative, etc. It follows easily that there are at least $N + 1$ points on the initial profile at which it has proper extrema. This contradiction demonstrates the theorem.

In order to summarize the preceding results we may illustrate them schematically in the following way, where the curved lines are velocity profiles.

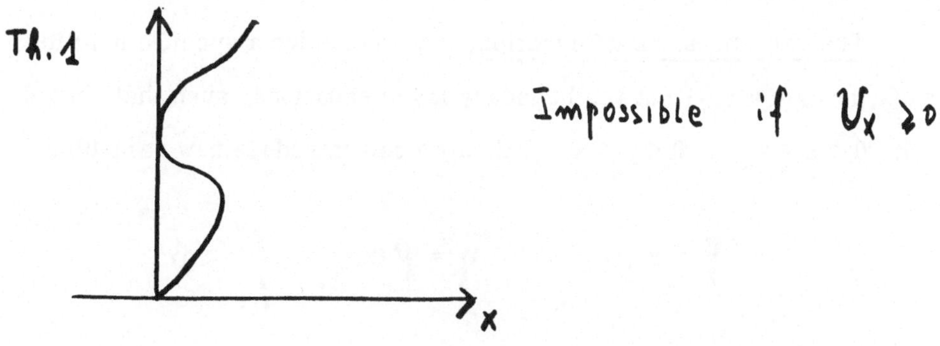

J.Serrin

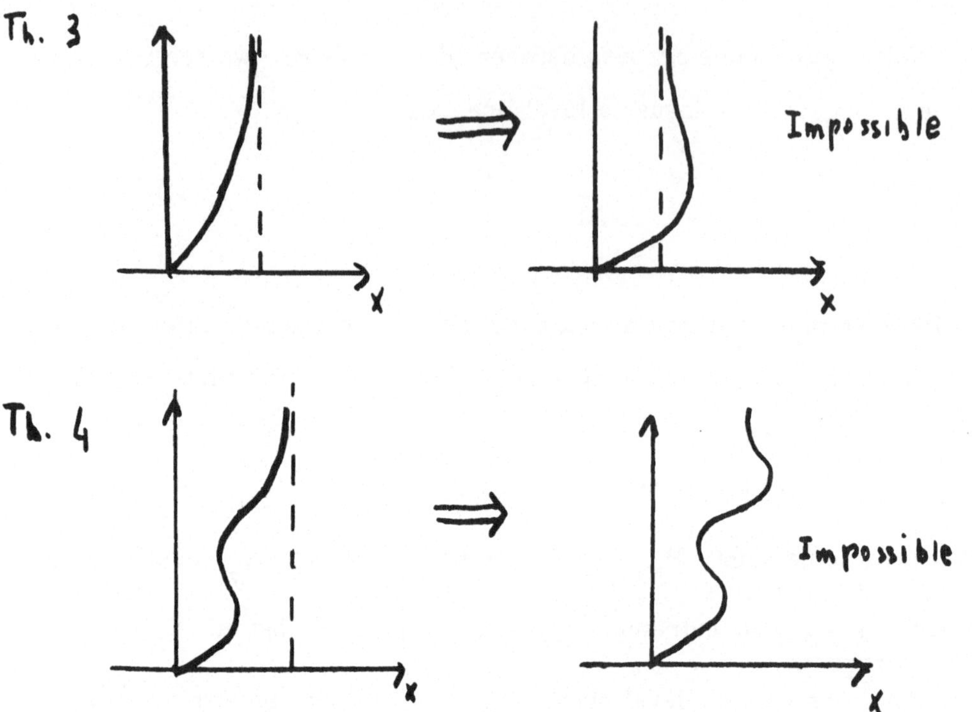

B. We continue our discussion of the Prandtl boundary layer theory by taking up the question of the uniqueness of the evolution problem. The proof here is based on the maximum principle and associated comparison arguments and involves, moreover, a preliminary change of variable due to von Mises. It is this change of variables which effectively limits the proof to the plane flow case. A consideration of three dimensional boundary layers here would be of the greatest interest, but has not yet been obtained.

<u>The von Mises transformation</u>. If we consider a specific solution $u = u(x, y)$, $v = v(x, y)$ of the boundary layer equations, such that $u > 0$ in R: $0 < x \leqslant x_o$, $0 < y < \infty$, then we can introduce new variables

$$\xi = x , \qquad \eta = \psi(x, y) = \int_0^y u \, dy ,$$

J. Serrin

where ψ is the streamfunction of the flow. The variables (ξ, η) range over the domain R' : $\left\{ 0 < \xi \leqslant x_0 , \quad 0 < \eta < \infty \right.$, and there is an obvious one-to-one connection between R and R' . The function $u(\xi, \eta) = u(x, y)$ satisfies, as is easily checked, the equation

$$(u^2)_\xi = (U^2)_\xi + \nu\, u(u^2)_{\eta\eta}$$

or equivalently

$$Z_\xi = \nu\, u Z_{\eta\eta} \quad , \quad Z = U^2 - u^2.$$

The analogy with the heat equation is apparent, ξ playing the role of time.[*]

THEOREM 1. If u is positive and bounded in R , then $\lim_{y \to \infty} u = U$ uniformly in x , $[6]$.

The proof uses a typical comparison argument. It is first of all evident[**] that our result is equivalent to

$$\lim_{\eta \to \infty} Z = 0 \qquad \text{uniformly in} \quad \xi \ (=t)$$

---

[*] For simplicity in writing further equations, we shall assume from here on that the kinematic viscosity $\nu = 1$. This clearly involves no loss of generality, since it can always be attained by a simple change of variables. The von Mises equations now take the form

$$(u^2)_\xi = (U^2)_\xi + u(u^2)_{\eta\eta} \quad ,$$

and

$$Z_\xi = u Z_{\eta\eta} \quad , \quad Z = V^2 - u^2.$$

[**] This requires a Heine-Borel type argument, which we omit.

J. Serrin

Now we know that $Z \to 0$ as $\eta \to \infty$ at $t = 0$. Thus for any $\varepsilon > 0$ we have

$$\left| Z(\eta, 0) \right| < \varepsilon \qquad \text{if} \quad \eta \geq \eta_0.$$

Now set

$$F = e^{-\frac{\alpha \eta^2}{t+1}} \, , \qquad \alpha = \text{constant}.$$

Then

$$F_\eta = -\frac{2\alpha}{t+1} F \, , \qquad F_{\eta\eta} = \left(\frac{4\alpha^2 \eta^2}{(t+1)^2} - \frac{2\alpha}{t+1}\right) F \, ,$$

and so, introducing a new linear differential operator $\mathcal{L}$,

$$\mathcal{L} F \equiv u F_{\eta\eta} - F_t = \left\{ u\left(\frac{4\alpha^2 \eta^2}{(t+1)^2} - \frac{2\alpha}{t+1}\right) - \frac{\eta^2}{(t+1)^2} \right\} F$$

$$= \left\{ \frac{\alpha \eta^2}{(t+1)^2} (4\alpha u - 1) - \frac{2\alpha u}{t+1} \right\} F \leq 0,$$

provided $\alpha = 1/4M$, where $M = \text{Max } u$. Similarly, set

$$G = \exp\left(\frac{\beta \eta^2}{1 - dt} + et\right)$$

so that

J.Serrin

$$\mathcal{L} G = \left\{ u\left(\frac{4\beta^2\eta^2}{(1-dt)^2} + \frac{2\beta}{1-dt}\right) - \left(\frac{d\beta\eta^2}{(1-dt)^2} + e\right) \right\} G$$

$$= \left\{ \frac{\beta\eta^2}{(1-dt)^2}(4\beta u - d) + \frac{2\beta u}{1-dt} - e \right\} G < 0$$

if

$$\beta = \frac{1}{4} , \qquad e = d = M , \qquad 0 < t < \frac{1}{2M} .$$

Finally set

$$H = \frac{\varepsilon + \gamma F + Z}{G} ,$$

where γ is a positive constant which we shall fix later. Since $\mathcal{L}(F+Z) \leqslant 0$ we find that

$$0 \geqslant \mathcal{L}(GH) = u(G_{\eta\eta} H + 2G_\eta H_\eta + GH_{\eta\eta}) - G_t H - GH_t$$

$$= G\mathcal{L}H + 2u\, G_\eta H_\eta + H\mathcal{L}G.$$

Finally, this may be written in the form

$$\mathcal{L}H + \left(\frac{2u\,G_\eta}{G}\right) H_\eta \leqslant \left(-\frac{\mathcal{L}G}{G}\right) H.$$

Now let us apply the maximum principle to this differential inequality. To this end, note that from the definition of H we have

J. Serrin

$$H \to 0 \quad \text{as} \quad \eta \to \infty \qquad \text{uniformly in } x .$$

Furthermore

$$H > 0 \quad \text{on} \quad t = 0, \qquad \eta \geqslant \eta_0$$

(since $|Z| < \varepsilon$), and

$$H > 0 \quad \text{on} \quad \eta = \eta_0 , \qquad 0 \leqslant t < 1/2M$$

(because $|Z| = |U^2 - u^2| \leqslant U^2 + u^2 \leqslant \gamma F$ if γ is suitably large). It follows from the maximum principle, since the coefficient of H is negative, that H cannot have a negative minimum. Hence $H > 0$. Therefore, we must have

$$\pm Z \leqslant \varepsilon + \gamma F.$$

But $F \to 0$ uniformly as $\eta \to \infty$ hence for all sufficiently large we have

$$|Z| < 2\varepsilon \qquad 0 \leqslant t < 1/2M.$$

Repeating the process a finite number of times proves the theorem.[*]

The preceding argument is a typical example of the application of comparison methods to determine the behavior of solutions of elliptic or parabolic equations at a singular point. Of course, it is clear that the application determines the choice of the comparison function, and that considerable care must occasionally be taken in finding a suitable comparison function. We co-

---

[*] The reader will observe that we have not used the condition $u = U$ at $y = \infty$, except at the initial line $x = 0$. It follows that the condition at infinity is superfluous provided $\lim\limits_{y \to \infty} u(y) = U(0)$.

J.Serrin

me now to the culminating theorem in our investigation of comparison methods in fluid mechanics, a proof that the time evolution of a velocity profile is uniquely determined.

THEOREM 2 . Let u and $\bar{u}$ be two solutions of the boundary layer equations corresponding to the same initial velocity profile $\tilde{u}(y)$. Suppose that u and $\bar{u}$ satisfy the hypothesis of Theorem 1 and that

$$u_{yy} < 0 \, ,$$

or more generally that

$$u_y(x, 0) > 0 \qquad , \qquad u_{yy} \leqslant B \, ,$$

for some constant B . Then $\bar{u} \equiv u.$

$\underline{Proof.}$ For simplicity we shall consider only the case when $u_{yy} < 0.$ This case includes most of the known exact solutions. We observe that since $u > 0, \ \bar{u} > 0,$ the von Mises transformation is applicable, and that by Theorem 1 both u and $\bar{u}$ tend uniformly to U as $\eta \to \infty$. Moreover, in the region $R' : \left\{ 0 < \xi \leqslant x_o , \quad 0 < \eta < \infty \right\}$ we have

$$(u^2)_\xi = (U^2)_\xi + u(u^2)_{\eta\eta} \quad , \quad (\bar{u}^2)_\xi = (U^2)_\xi + \bar{u}(\bar{u}^2)_{\eta\eta} \ .$$

Now set $\phi = \bar{u}^2 - u^2$, whence by subtraction of the preceding equations there arises

$$(5) \qquad \begin{aligned} \phi_\xi &= \bar{u}(\bar{u}^2)_{\eta\eta} - u(u^2)_{\eta\eta} \\ &= \bar{u}\,\phi_{\eta\eta} + (\bar{u} - u)\,(u^2)_{\eta\eta} \\ &= \bar{u}\,\phi_{\eta\eta} + a\,\phi \ , \end{aligned}$$

89

J. Serrin

where

$$a = \frac{(u^2)_{\eta\eta}}{u + \bar{u}} = \frac{2u_{yy}}{u(u + \bar{u})} < 0 .$$

Now obviously

$$\phi \to 0 \quad \text{as} \quad \eta \to \infty, \text{ uniformly in } \xi$$

$$\phi = 0 \quad \text{when} \quad \xi = 0 \quad \text{and when} \quad \eta = 0.$$

From equation 5 it is clear that ϕ cannot have either a positive maximum or a negative minimum in R', hence $\phi = 0$. Thus

$$u(\xi, \eta) \equiv \bar{u}(\xi, \eta).$$

To conclude the proof, note that the inverse von Mises transformation is given by

$$y = \int_0^\eta \frac{d\eta}{u(\xi, \eta)} , \quad x = \xi .$$

Hence for a given value of η, the corresponding values y and $\bar{y}$ for the two solutions are the same. Consequently,

$$u(x, y) = u(\xi, \eta) = \bar{u}(\xi, \eta) = \bar{u}(x, y).$$

This complets the proof.

As we have already observed, an argument which serves to prove a uniqueness theorem can frequently be used to obtain a corresponding <u>comparison theorem</u>. This is the case here, with the following result.

THEOREM 3. Let u and $\bar{u}$ be solutions of the boundary layer equa-

J. Serrin

tions corresponding to initial profiles $\tilde{u}$ and $\tilde{\bar{u}}$, and streaming velocities U and $\bar{U}$. Suppose that u and $\bar{u}$ satisfy the hypotheses of Theorem 1 and that <u>either</u> $u_{yy} < 0$ or $\bar{u}_{yy} < 0$. Suppose also that, in the von Mises variables,

$$\tilde{u}(\eta) \leqslant \tilde{\bar{u}}(\eta)$$

and that

$$(\bar{U}^2 - U^2)_x \geqslant 0.$$

Then

$$u(x, y) \leqslant \bar{u}(x, y).$$

The proof is almost the same as before except that we are led to the equation (Supposing $u_{yy} < 0$)

$$(5') \qquad \phi_\xi = \bar{u}\,\phi_{\eta\eta} + a\,\phi + (\bar{U}^2 - U^2)_\xi \quad , \qquad\qquad (\phi = \bar{u}^2 - u^2).$$

In addition, ϕ satisfies the boundary conditions

$$\phi = \bar{U}^2 - U^2 \geqslant 0 \qquad \text{at infinity}$$

$$\phi = \tilde{\bar{u}}^2 - \tilde{u}^2 > 0 \qquad \text{at } \xi = 0$$

$$\phi = 0 \qquad \text{at } \eta = 0 \quad .$$

Consequently, since ϕ cannot have a negative minimum,[*] we have $\phi \geqslant 0$ and

$$u(\xi, \eta) \leqslant \bar{u}(\xi, \eta).$$

To conclude the proof, we have for any fixed value of η ,

---

[*] At such a point, we would have $\bar{u}\,\phi_{\eta\eta} - \phi_\xi < 0.$

J. Serrin

$$y = \int_0^{\eta} \frac{d\eta}{u(\xi,\eta)} \geq \int_0^{\eta} \frac{d\eta}{\bar{u}(\xi,\eta)} = \bar{y} .$$

Thus it is evident that

$$(6) \qquad u(x,y) = u(\xi,\eta) \leq \bar{u}(\xi,\eta) = \bar{u}(x,\bar{y}) .$$

Now we have $u_y > 0$ [*], hence $u(x,\bar{y}) \leq u(x,y)$, and the required inequality is proved. If $\bar{u}_{yy} < 0$, we prove in almost the same way that (6) holds. The final result the follows since $\bar{u}_y > 0$ and $\bar{u}(x,\bar{y}) \leq \bar{u}(x,y)$.

Remark. This theorem is analogous to certain results of Nickel[2], but neither contains his results nor is contained in them. Let us note several consequences which have some physical interest.

1. Suppose the external speed is constant , $U \equiv U_0$. Let $\bar{u}$ be the corresponding Blasius solution

$$\bar{u} = U_0 \, f' \left(\frac{y}{\sqrt{2\nu x / U_0}} \right) .$$

Then for any other solution u such that $\tilde{u} \leq U_0$ we have

$$u(x,y) \leq U_0 \, f' \left(\frac{y}{\sqrt{2\nu x / U_0}} \right) .$$

2. Suppose the external flow is of the form $U = Cx^m$, $m > 0$. Suppose also that initially

$$0 \leq \tilde{u} \leq U(0) ,$$

---

[*] Since $u_{yy} < 0$, and $u = 0$ at $y = 0$, $u = U$ at $y = \infty$, one sees that necessarily $u_y > 0$ for $0 < y < \infty$.

J. Serrin

ie. $\tilde{u} \not\equiv 0$. Then any solution u must necessarily coincide with the well-known Falkner-Skan solution, for this latter is a particular solution with $\tilde{u} = 0$.

3. If we examine the so-called outer boundary condition $u = U$ at $y = \infty$, it must be admitted that this is only an artifice to obtain a solution with the more realistic behavior

$$u \approx U \qquad \text{at} \quad y = 0(\sqrt{\nu}) \quad .$$
$$u_y \not\approx 0$$

Let us now consider the boundary layer equations with this (more realistic) boundary condition. By the change of variables $y = \sqrt{\nu}\,\bar{y}, \quad \tilde{v} = \sqrt{\nu}\,v, \quad \bar{u} = u$ our problem becomes

$$\bar{u}\bar{u}_x + \bar{v}\bar{u}_{\bar{y}} = U U_x + \bar{u}_{\bar{y}\bar{y}}$$

$$\bar{u}_x + \bar{v}_{\bar{y}} = 0$$

with conditions

$$\bar{u} = U \qquad \text{at} \quad \bar{y} = 0(1) \quad .$$
$$\bar{u}_{\bar{y}} = 0$$

Since both these boundary conditions cannot be simultaneously met, let us keep just the first, and consider solutions of the boundary layer equations such that

$$\bar{u} = U \qquad \text{on a curve} \quad \bar{y} = f(x) \quad .$$

To compare this with a solution of the usual problem

$$u = U \qquad \text{on} \quad \bar{y} = \infty ,$$

J. Serrin

we use the ideas of the preceding theorem. In particular, let us suppose that $u_{yy} < 0$ for the solution u , and that the initial conditions are such that

$$\tilde{u}(\eta) \leqslant \tilde{\bar{u}}(\eta).$$

In particular, this will be the case if the initial conditions for both u and the solution $\bar{u}$ are identical. Then applying the methods of Theorem 3 one is easily led to the inequality

$$u(x, y) \leqslant \bar{u}(x, y),$$

which is illustrated below.

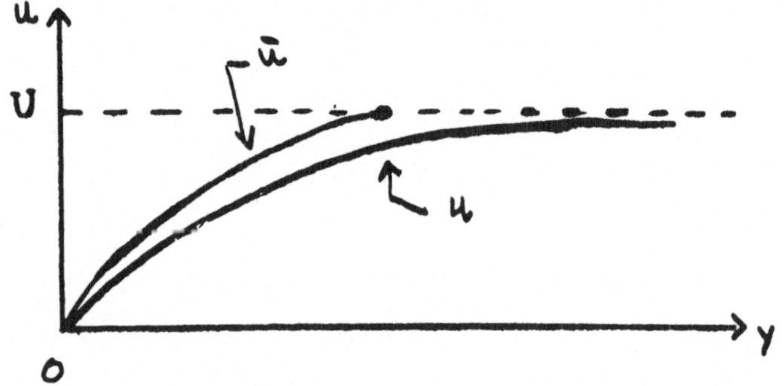

·Although there is no compelling reason that $\bar{u}$ need be close to u , the graph does indicate that in order to guarantee $\bar{u}_y \not\geqslant 0$, we should take the outer edge of the boundary layer in a region where $u \approx U$. If this is done, however, then $\bar{u}$ differs only slightly from u , and we might as well take u itself for the boundary layer profile. Thus we see rather clearly the <u>reason</u> for the success of Prandtl's artifice.

J. Serrin

REFERENCES : CHAPTER III

[1] K. Nickel, Einige Eigenschaften von Lösungen der Prandtlschen Grenzschichtdifferentialgleichungen. Arch. Rat. Mech. Anal. $\underline{2}$ (1958), 1-31.

[2] K. Nickel, Eine einfache Abschätzung für Grenzschichten. Ingenieur Archiv, $\underline{31}$ (1962), 85-100.

[3] K. Nickel, Ein Eindeutigkeitssatz für instationäre Grenzschichten. Math. Zeit. $\underline{74}$ (1960), 209-220.

[4] W. Velte, Eine Anwendungen des Nirenbergschen Maximumprinzips für parabolische Differentialgleichungen in der Grenzschichttheorie. Arch. Rat. Mech. Anal. $\underline{5}$ (1960), 420-431.

[5] H. Görtler, Über die Lösung nichtlinearen partieller Differentialgleichungen vom Riebungsschichttypus. ZAMM $\underline{30}$ (1950), 265-267.

[6] J. Serrin, Mathematical Aspects of Boundary Layer Theory. Univ. of Minnesota, 1963.

J. Serrin

IV. STRESS, VORTICITY, AND ENERGY AVERAGES

A second classical method for studying the behavior of a mechanical system is through various averaging procedures, in which the primary interest is in the space or space-time average of some physical variable. These methods are valuable because they allow us to concentrate on overall behavior, rather than on unimportant local variations. Statistical mechanics is perhaps the best example which can be offered of the extreme importance of the averaging idea, but even so, averaging can play an important part in field theories and in continuum mechanics, and it is this side of the picture which we discuss here. It should be added that even if mathematicians were not gifted with physical insight, they would still be led to the idea of averaging, for as we shall see, it is an extremely natural process at the elementary level; while at the advanced level it allows one to bring into play the powerful techniques of functional analysis.

A. Let us begin our discussion by setting down, in differential form, the fundamental equations of motion of a continuous medium, namely

$$\rho\,(\vec{a} - \vec{f}) = \operatorname{div} \vec{T}\ ,$$

where $\vec{T}$ is the stress tensor, assumed to be symmetric, $\vec{a}$ is the acceleration, and $\vec{f}$ the external force. As with the special case of the Navier-Stokes equations, I assume that you are also familiar with this equation. For the reasons indicated earlier, it is natural to multiply this equation by a weighting function or vector ϕ and average over a (possibly moving) volume Ω, thus

$$\int_{\Omega} \rho\,\phi\,(\vec{a} - \vec{f})\,dv = \int_{\Omega} \phi\,\operatorname{div}\vec{T}\,dv.$$

J.Serrin

The right hand side may be rewritten in the form

$$\int_\Omega (\mathrm{grad}\,\phi)\cdot\vec{T}\,dv + \oint_{\partial\Omega}\phi\;\vec{n}\cdot\vec{T}\,ds.$$

Combining the two proceding equations then leads to the important formula

$$(7)\qquad \int\left[\rho\,\phi\,(\vec{a}-\vec{f}) + (\mathrm{grad}\,\phi)\cdot\vec{T}\right] = \oint\phi\,\vec{t}$$

where $\vec{t}$ is the stress vector on the surface. Here for simplicity we have omittes the conventional infinitesimals dv and ds , as well as the symbols Ω and $\partial\Omega$ denoting the set of integration. These things will be apparent from the contest in all future formulas.

The standard averaging technique which leads to formula (7) can be summarized by the following steps :

1) Multiply both sides of the basic equation in question by a test (or weight) function ϕ .

2) Integrate over an appropriate region Ω

3) Simplify the result by using the divergence theorem (i. e. integration by parts)

4) Choose an appropriate test function ϕ .

We intend to illustrate the last step by several particular choices of the function ϕ . It will be convenient, however, first to recall a simple result, the so-called .

TRANSPORT THEOREM. Let Ω be a (possibly moving) volume in the interior of a region of fluid motion, and let   F   denote some physical variable. Then

J. Serrin

$$\frac{d}{dt} \int \rho\, f = \int \rho\, \frac{df}{dt} - \oint \rho\, f\, (G - \vec{v} \cdot \vec{n})\, ds,$$

where G denotes the normal outward speed of the boundary of Ω. In particular, if Ω is a volume V moving with the fluid, then

$$\frac{d}{dt} \int \rho\, f = \int \rho\, \frac{df}{dt} \quad .$$

The transport theorem in the form noted above is proved in $\left[10\right]$, as well as in articles in the Handbuch der Physik by Serrin and by Toupin and Truesdell.

Let us now obtain our first application of (7) by setting $\phi = \vec{v}$, and assuming that $\Omega = V$ is a volume moving with the fluid. Since

$$\vec{v} \cdot \vec{a} = \frac{1}{2} \frac{d}{dt} q^2 \, , \qquad \text{grad } \vec{v} : T = T : D$$

where grad $\vec{v}$ is the tensor $\partial v_i / \partial x_j$, this gives

$$(8) \qquad \frac{d}{dt} \int \frac{1}{2} \rho\, q^2 = \oint \vec{v} \cdot \vec{t} + \int \left[\rho \vec{v} \cdot \vec{f} - \vec{T} : \vec{D} \right] \, ,$$

the important energy transfer formula. We will return to the notion of energy averages later from a slightly different standpoint.

A second example arises if one sets $\phi = \vec{x}$ (the position vector) in order to examine the <u>first moments</u> of the Cauchy equation. Then grad ϕ is the identity matrix, and we obtain the interesting formula

$$\int \left[\rho\, \vec{x}\, (\vec{a} - \vec{f}) + \vec{T} \right] = \oint \vec{x}\, \vec{t}.$$

J. Serrin

Signorini has pointed out several applications to the <u>static case</u>, when the preceding formula reduces to the simple stress average identity

$$\int \vec{T} = \oint \vec{x}\,\vec{t} + \int \rho\,\vec{x}\,\vec{f}.$$

Thus in particular <u>the stress average is completely determinate from the external load.</u>

To take a third example, suppose ϕ is the dyadic tensor $\vec{x}\,\vec{x}$. Confining our attention to the static case $\vec{a} = 0$, and using tensor notation, the result is

$$\int (x_k\,x_\ell)_{,i}\,T_{ij} = \oint x_k\,x_\ell\,t_i + \int \rho\,x_k\,x_\ell\,f_i = a_{ik\ell} \quad ,$$

the $a_{ik\ell}$ being a constant tensor which is determined by the given external loading. Now we have

$$(x_k\,x_\ell)_{,i}\,T_{ij} = T_{ij}\,\delta_{ki}\,x_\ell + T_{ij}\,x_k\,\delta_{\ell i} = T_{kj}\,x_\ell + T_{\ell j}\,x_k \quad .$$

By appropriate permutation of the indices, and combination of the results, we then get the remarkable formula

$$\int x_i\,T_{k\ell} = b_{ik\ell} = \frac{1}{2}\,(a_{k\ell i} + a_{\ell ki} - a_{ik\ell}) \quad ,$$

that is, <u>the first moments of the stress are likewise determined by the external loading.</u> The preceding two italicized results imply that <u>the four quantities</u>

(9)
$$\int \vec{T} \;, \qquad \int x_1\,\vec{T} \;, \qquad \int x_2\,\vec{T} \;, \qquad \int x_3\,\vec{T}$$

J. Serrin

are determinate from the loading. Now the functions

$$1, \; x_1, \; x_2, \; x_3$$

are orthogonal over V , provided that the origin is chosen at the centroid of V and the axes are oriented in the principal directions. Assuming this to be the case we can apply Bessel's inequality to obtain bounds for the square of individual stress components in terms of the loading. Indeed the quantities (9) are just Fourier coefficients of T , which we denote by $b_0, \ldots, b_3$. Thus if τ is some particular stress component we have

$$\int_V \tau^2 \, dx \geqslant \sum_0^3 \frac{b_i^2}{\lambda_i}$$

where

$$\lambda_o = \int 1 \, dv = V \quad , \qquad \lambda_i = \int x_i^2 \, dv \; .$$

This lower bound for internal stresses is due to Signorini, who gives a number of examples. Grioli has similarly investigated higher mements.

We indicate here, without any particular motivation, the more important known vorticity average theorems.

$$1) \qquad \int \omega^2 = \int \text{grad } \vec{v} : \text{grad } \vec{v} - \oint \vec{a} \cdot \vec{n} = 2 \int \vec{D} : \vec{D} - 2 \oint \vec{a} \cdot \vec{n} \; ,$$

$$2) \qquad \int (\vec{\omega} \times \vec{v}) = \oint \left[\vec{v}(\vec{v} \cdot \vec{n}) - \frac{1}{2} q^2 \vec{n} \right] \; , \qquad \bullet$$

$$3) \qquad \frac{d}{dt} \int \vec{\omega} \cdot \text{grad } \phi = \oint \left[\vec{\omega} \cdot \vec{n} \frac{d\phi}{dt} - (\vec{a} \times \vec{n}) \cdot \text{grad } \phi \right] \; ,$$

where $\text{div } \vec{v} = 0$ in 1) and 2). These are all essentially kinematic. The second one is due to Lamb ; the third is basically due to Ertel. (For proof

100

J. Serrin

of the above results, see $\left[5\right]$).

At a fixed wall $\vec{a} \cdot \vec{n} = 0$, and if the fluid adheres at the wall then $\vec{v}, \vec{a}$, and $\vec{\omega} \cdot \vec{n}$ are also zero. In these cases, then, the surface integrals in the above identities all vanish, resulting in extremely simple formulas. For example, if we take $\phi = \vec{x}$ in the last formula, and assume that the motion takes place within a rigidly bounded region V we obtain the remarkable theorem that

$$\int_V \vec{\omega} \equiv \text{constant} .$$

Similarly if $\phi = S = $ entropy, and we consider non-conducting, inviscid fluids, there results

$$\int_V \vec{\omega} \cdot \text{grad } S \equiv \text{constant, following motion.}$$

$\underline{B}.$ We have not yet turned to the energy average, which is in many respects the most interesting and most useful, especially when used to estimate the energy of a difference motion. We shall treat viscous incompressible fluids first, then insert a section on compressible fluids, and finally return to incompressible fluids, where we shall discuss the general theory of the initial value problem.

Our interest here will be in the application of averaging methods to the initial value problem, and in particular to the corresponding questions of uniqueness and stability. Consider a bounded region $\Omega = \Omega(t)$ occupied by a viscous incompressible fluid, with prescribed velocity distribution on the boundary $\partial\Omega$. In the case of greatest interest Ω is bounded by (possibly moving) material walls and the boundary conditions arise from the adherence condition at the walls. We now ask, is the fluid motion under these circumstances uniquely determined by the initial velocity distribution $\vec{v}_0(x)$,

101

J. Serrin

and if so, is the motion stable with respect to perturbations of the initial state? The basic technique in this study is an identity expressing the rate of change of total energy.

In studying this and related problems, it is convenient to begin with the
Navier-Stokes equations in appropriately averaged form. Let $\mathcal{D}$ denote the
linear space of vector functions $\vec{\phi} = \vec{\phi}(x, t)$ which are divergence free in the
basic region Ω and vanish on the boundary of Ω. Taking the Navier-Stokes equation in the form

$$\vec{a} - \vec{f} = -\text{grad } P/\varrho + \nu \Delta v \, ,$$

multiplying through by $\vec{\phi}$ and forming the scalar product, and finally averaging over Ω, we get

$$(10) \qquad \int \left[\vec{\phi} \cdot (\vec{a} - \vec{f}) + \nu \, \text{grad } \vec{\phi} : \text{grad } \vec{v} \right] = 0 \qquad \vec{\phi} \in \mathcal{D},$$

which is, of course, nothing more than our original identity expressed for the
Navier-Stokes equation.

Now let v and $\tilde{v}$ be two possibly different velocity fields in Ω,
each satisfying the given boundary conditions. Let $u = v - \tilde{v}$ be the perturbation velocity field,* and

$$K = K(t) = \frac{1}{2} \int u^2$$

the perturbation energy over Ω. Then we have the important formula

$$(11) \qquad \frac{dK}{dt} = \int (u \cdot \text{grad } u \cdot v - \nu \, \text{grad } u : \text{grad } u).$$

---

* From here on, we shall generally omit writing the arrow over a vector quantity.

J.Serrin

<u>Proof.</u> Writing (10) also for $\tilde{v}$, we have

$$\int \left[\phi \cdot (\tilde{a} - \tilde{f}) + \nu \,\mathrm{grad}\, \phi : \mathrm{grad}\, \tilde{v} \right] = 0$$

Hence by subtraction

$$\int \left[\phi \cdot (\tilde{a} - a) + \nu \,\mathrm{grad}\, \phi : \mathrm{grad}\, u \right] = 0.$$

Now suppose that $\phi = u$. Then one shows easily that

$$\int u \cdot (\tilde{a} - a) = \int \left(\frac{1}{2} \frac{\partial u^2}{\partial t} - u \cdot \mathrm{grad}\, u \cdot v \right) ,$$

and the required result follows at once, using the transport theorem.

THEOREM 1. Let v and $\tilde{v}$ be two continuously differentiable solutions of the Navier-Stokes equations in Ω , both assuming the given boundary data. Then

$$K \leqslant K_0 \exp(V^2 - \alpha \nu^2/d^2)\, t/\nu .$$

Here $K_0 = K(0)$ is the initial perturbation energy, V is the maximum of the speed $|v|$ of the basic flow in the time interval $(0,t)$, d is the diameter of a sphere containing Ω , and α is a pure number,

$$\alpha \approx 80 ,$$

<u>Remarks.</u> Before proving this result we observe that it implies simultaneously a uniqueness theorem, and a stability theorem.

103

J. Serrin

1) If the flows v and $\tilde{v}$ have the same initial data, then obviously $K_0 = 0$. Consequently $K = 0$, and so

$$u = v - \tilde{v} = 0 ,$$

that is, the two flows must be identical for all $t \geqslant 0$. This was first proved by E. Foà in 1930.

2) Suppose that in the time interval $(0, \infty)$ we have

$$Re = \frac{dV}{\gamma} < \sqrt{80} \approx 8,98$$

then

$$K \leq K_0 \, e^{-\beta t} \qquad\qquad \text{where } \beta > 0 .$$

Consequently $K \to 0$ as $t \to \infty$ and the basic motion v is stable in the mean with respect to arbitrary disturbances in the initial data. We have called 8.98 the Reynolds number for universal stability.[*]

3) Small change in data causes small changes in the solution; the problem is thus well-set in the sense of Hadamard (C-H, pp. 226 ff.)

Proof of theorem. This hinges on the inequality

$$\alpha \, d^{-2} \int u^2 \leq \int \mathrm{grad}\ u: \mathrm{grad}\ u , \qquad\qquad \alpha \approx 80 ,$$

which was recently demonstrated by Payne and Weinberger. Although we shall assume this inequality without proof, it should be clear that it holds for some

---

[*] It should be emphasized here that the stability here is with respect to arbitrary disturbances, whereas in the usual linearized theory one obtains stability only with respect to infinitesimal disturbances.

J.Serrin

$\alpha > 0$, according to the well known fact that one can estimate the L_2 norm of a function in terms of the L_2 norm of its first derivatives (Poincaré's inequality). Now it is clear from the Cauchy inequality that

$$u \cdot A \cdot v \leqslant \frac{1}{2} \; (A : A + u^2 v^2) \; ,$$

hence we obtain easily from (11)

$$\frac{dK}{dt} \leqslant \frac{1}{2\nu} \int (u^2 v^2 - \nu^2 \; \text{grad } u : \text{grad } u)$$

$$\leqslant \frac{1}{2\nu} \int (u^2 v^2 - \alpha \nu^2 d^{-2} u^2)$$

$$- \frac{1}{\nu} (V^2 - \alpha \nu^2 / d^2) \, K \; .$$

Integrating this differential inequality yields the required estimate.

I observed in a paper several years ago that the stability of the basic flow v could be reduced to the variational problem of determining the least number $\tilde{\nu}$ such that

$$\int (\tilde{\nu} \; \text{grad } u : \text{grad } u - u \cdot \text{grad } u \cdot v) \geqslant 0$$

for all divergence free vector fields u which vanish on the boundary of Ω . Since the integral is homogeneous in u , and since

$$\int u \cdot \text{grad } u \cdot v = - \int u \cdot D \cdot u \; ,$$

where D is the rate of deformation tensor, we can obviously reformulate

J. Serrin

the problem as that of determining the maximum of the integral

$$- \int u \cdot D \cdot u$$

subjedt to the side conditions

$$\text{div } u = 0, \qquad\qquad \int \text{grad } u : \text{grad } u = 1.$$

The corresponding Euler-Lagrange equations for this variational problem have the remarkably simple form

$$(12) \qquad \begin{aligned} & u \cdot D = - \text{grad } \lambda + \gamma^* \Delta u \\ & \text{div } u = 0, \qquad\qquad u = 0 \text{ on } \partial\Omega. \end{aligned}$$

Here $\lambda = \lambda(x)$ and γ^* are the "eigenvalues" of the problem. We now have the basic

THEOREM 2. The motion is stable if $\gamma > \tilde{\gamma}$, where $\tilde{\gamma}$ is the greatest eigenvalue of the problem (12).

Proof. By standard methods of the calculus of variations we know there is a maximizing vector $\tilde{u}$ for the integral $- \int u \cdot D \cdot u$, and that $\tilde{u}$ is a solution of (12). Let $\tilde{\gamma}$ be the eigenvalue corresponding to $\tilde{u}$. We show first that $\tilde{\gamma}$ is the greatest eigenvalue.

Indeed let u be any other solution of (12) with eigenvalue γ^* , and suppose u (as well as $\tilde{u}$) normalized so that $\int \text{grad } u : \text{grad } u = 1$. Then

$$- \int u \cdot D \cdot u = \int u \cdot (\text{grad } \lambda - \gamma^* \Delta u) = \gamma^* \int \text{grad } u : \text{grad } u = \gamma^*.$$

Similarly we find

$$- \int u \cdot D \cdot u = \tilde{\gamma}.$$

J. Serrin

Hence since $\tilde{u}$ solves the variational problem, we have $\tilde{\nu} \geqslant \nu^{*}$, and $\tilde{\nu}$ is the greatest eigenvalue.

Now we shall show that the motion is stable if $\nu > \tilde{\nu}$. But in this case, since $\tilde{\nu}$ is the maximum of $-\int u \cdot D \cdot u$ under the side conditions, it is clear that

$$\int (\nu \operatorname{grad} u : \operatorname{grad} u - u \cdot D \cdot u) > 0.$$

The motion is therefore stable if $\nu > \tilde{\nu}$.

If one wishes to consider the stability of a specific flow, it is clear that the variational approach will provide in general a better criterion than that of Theorem 1. The only flow for which this program has been carried out, however, is that of Couette motion between circular cylinders, and even here the results are fragmentary. Further research into this problem, especially with regard to the dependence of the stability on the spacing between the cylinders woold seen to me of the greatest importance.

The preceding uniqueness theorem (c.f. Remark 1) applies only to a bounded region Ω. There is a corresponding result for an unbounded region which is available under either one of the following hypotheses :

I. The velocity v is uniformly bounded for $0 \leqslant t \leqslant T$, and also is contained in the set V (cf. Chap. V).

II. The quantities v, and grad v, are uniformly bounded for $0 \leqslant t \leqslant T$, and in addition the pressure p tends to a limit p_o at infinitily in such a way that

$$\left| p - p_o \right| \leqslant \text{Const.} \ |x|^{-\frac{1}{2} - \varepsilon}$$

where A and ε are fixed positive constants.

More precisely, we have the following theorem: Jet $\Omega = \Omega(t)$ deno-

J. Serrin

te the <u>exterior</u> of a bounded region in space. Let v and $\tilde{v}$ be two solutions of the Navier-Stokes equations in Ω , having the same initial data and the same boundary data on $\partial\Omega$. Suppose also that v and $\tilde{v}$ satisfy <u>either</u> hypothesis I or hypothesis II. Then $v \equiv \tilde{v}$ for $o \leq t \leq T$.

This theorem, whose proof we omit, is due to Leray (Hypothesis I) and to Graffi (Hypothesis II).[*]

<u>C.</u> We shall conclude our discussion of energy averages by considering the uniqueness and stability of the initial value problem for a compressible inviscid fluid. Except that the physical properties of the fluid are now quite altered, this is precisely the problem which we have just treated. Moreover, the mathematical techniques involved in the present discussion are essentially the same as before, though the proofs do take a more complicated appearance. The <u>important</u> point to be made is that the <u>method</u> applies almost equally well, in spite of the added physical complications.

The appropriate differential equations governing the motion are well known, namely

$$(13) \quad \begin{cases} \dfrac{d\varrho}{dt} + \varrho \ \mathrm{div} \ v = 0, \\[2ex] \varrho \ (a - f) + \mathrm{grad} \ p = 0, \\[2ex] \dfrac{dS}{dt} = 0, \ (S = \text{entropy}) \end{cases}$$

together with an equation of state $p = p (\varrho, S)$. We assume that the density ϱ is always <u>positive</u>, and that

---

[*] It is my opinion that an even stronger result should hold, probably requiring no hypotheses either on $\mathrm{grad} \ v$ or on p . For a similar result concerning parabolic equations, one may consult a paper of Rosenbloom in Contributions to Partial Differential Equations, Princeton, 1953

J. Serrin

$$\left(\frac{\partial p}{\partial \varrho}\right)_S > 0 \; ,$$

corresponding to a <u>real</u> speed of sound in the fluid. Now suppose that

$\varrho = \varrho\,(x, t), \quad v = v(x, t), \quad S = S(s, t)$ is a given solution of the above equations for $\; t \geq 0$. It is well known that the system (13) is hyperbolic. For this reason, if $\; K \;$ denotes a come-like region in space time, with cross section $\Omega(t)$, such that the mantle (lateral boundary) of $\; K \;$ is a <u>characteristic sur-</u><u>face</u>, we expect that any other solution

$$\tilde{\varrho} \; , \; \tilde{S} \; , \; \tilde{v}$$

of the flow equations which takes the same initial values on $\; \Omega(0) \;$, will be identical with the original solution. The situation is shown, for two space dimensions, in the accompanying figure. The characteristic condition at the boundary of $\; \Omega(t) \;$ is $\; v \cdot n = G + c \; , \;$ where $\; G \;$ denotes the outward

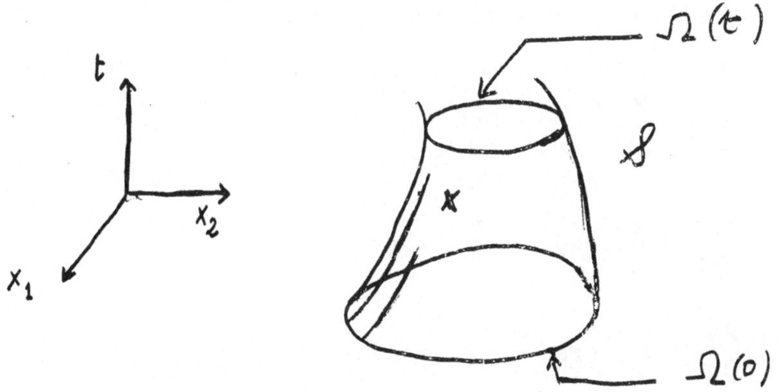

normal velocity of the boundary, thinking of this as a moving surface in space. Thus the boundary of $\; \Omega \;$ has the property that at each point the fluid in the basic motion ϱ , S, v is leaving $\; \Omega \;$ with <u>precisely</u> sonic speed with respect to the velocity of the boundary.

109

J. Serrin

THEOREM. Consider a solution ϱ, S, v of the system (13) in a region K of space-time bounded by the plane $t = 0$ and a mantle $\mathcal{S}$ on which $v \cdot n = G + c$. Suppose that $\tilde{\varrho}, \tilde{S}, \tilde{v}$ is another solution in K which assumes for $t = 0$ the same values as the first one. Then the two solutions are identical in K, [14].

We may remark that this theorem is the analogue of the well known uniqueness theorem for the wave equation, where the solution is determined by its initial values in a certain retrograde light cone. In fact, this latter theorem is also proved by an application of the energy averaging method, albeit an extremely simple one, cf. Courant-Hilbert, Vol. II, pp. 642-646. In this connection I should also point out the large body of work on symmetric linear hyperbolic systems initiated by the papers of K. O. Friedrichs, where the idea of energy identities again plays a crucial role. I believe that some of this material was covered in the CIME lecture of R. S. Phillips earlier this summer. Although this work has considerable generality it must be noted that it does not aplly to the present case, since the equations (13) are <u>neither</u> symmetric nor linear.

We may now turn to the proof of the theorem. We observe, to begin with, that the alternate flow $\tilde{\varrho}, \tilde{S}, \tilde{v}$ satisfies (13), that is

$$\frac{\partial \tilde{\varrho}}{\partial t} + \tilde{v} \; \text{grad} \; \tilde{\varrho} + \tilde{\varrho} \; \text{div} \; \tilde{v} = 0$$

$$\tilde{\varrho} \left(\frac{\partial \tilde{v}}{\partial t} + \tilde{v} \cdot \text{grad} \; \tilde{v} - f \right) + \text{grad} \; \tilde{p} = 0$$

$$\frac{\partial \tilde{S}}{\partial t} + \tilde{v} \cdot \text{grad} \; \tilde{S} = 0.$$

Let us subtract from each of these the corresponding equation for the basic flow, multiply the resulting equations respectively by

110

J. Serrin

$$\varrho' = \tilde{\varrho} - \varrho \, , \qquad v' = v - v, \qquad S' = \tilde{S} - S,$$

and integrate over a cross section $\Omega(t)$. (In essence, this is the procedure already successfully applied in the incompressible case.) The resulting equations are unfortunately fairly complicated, so we shall treat in detail only the last one. In this case, subtraction yields

$$\frac{\partial S'}{\partial t} + v \cdot \text{grad } S' + v' \cdot \text{grad } S = 0.$$

Multiplying by S' then gives

$$\frac{1}{2} \left(\frac{\partial S'^2}{\partial t} + v \cdot \text{grad } S'^2 \right) + S'v' \cdot \text{grad } S = 0$$

or simply

$$\frac{dS'^2}{dt} + 2S'v' \cdot \text{grad } S = 0 \, , \quad \left(\frac{d}{dt} = \frac{\partial}{\partial t} + v \cdot \text{grad} \right).$$

Multiplying both sides by ϱ and <u>integrating over $\Omega(t)$</u> then gives

$$\frac{d}{dt} \int \varrho S'^2 = - \int 2\varrho S'v' \cdot \text{grad } S - \oint \varrho (v \cdot n - G)S'^2$$

where we have used the transport theorem in the form

$$\frac{d}{dt} \int \varrho f = \int \varrho \frac{df}{dt} - \oint \varrho (v \cdot n - G)f \, .$$

But

$$2\varrho S'v' \cdot \text{grad } S \leq \text{const.}(v'^2 + S'^2),$$

whence with the help of the given boundary conditions,

111

J. Serrin

$$\frac{d}{dt}\int \varrho\, S'^2 \leq \text{Const.} \quad (v'^2 + S'^2) - \quad c\, S'^2$$

Carrying out the same procedure for the equations of motion, and incidentally using the continuity equations at a crucial point, yields

$$\frac{d}{dt}\int \varrho\,(v'^2 + \frac{c^2}{\varrho^2}\,\varrho'^2 + \frac{2\alpha}{\varrho^2}\,\varrho'S') \leq \text{Const.} \int (\varrho'^2 + v'^2 + S'^2)$$

$$- \oint \left[\varrho\, c(v'^2 + \frac{c^2}{\varrho^2}\,\varrho'^2 + \frac{2\alpha}{\varrho^2}\,\varrho'S') + 2(c^2\varrho' + \alpha\, S')v'\bullet n \right],$$

where $\alpha = (\partial p/\partial S)_\varrho$ is a thermodynamic variable. Now multiply the S' equation by a large constant A and add it to the one just written, to obtain

$$\frac{d}{dt}\int \varrho\,(v'^2 + \frac{c^2}{\varrho^2}\,\varrho'^2 + \frac{2\alpha}{\varrho^2}\,\varrho'S' + AS'^2)$$

$$\leq \text{Const.} \quad (\varrho'^2 + v'^2 + S'^2)$$

$$- \oint \left[\varrho\, c(v'^2 + \frac{c^2}{\varrho^2}\,\varrho'^2 + \frac{2\alpha}{\varrho^2}\,\varrho'S' + AS'^2) + 2(c^2\varrho' + \alpha S')v'\bullet n \right.$$

where the Const. is perhaps different than before, but a constant nonetheless. Define

$$J = (v'^2 + \frac{c^2}{\varrho^2}\,\varrho'^2 + \frac{2\alpha}{\varrho^2}\,\varrho'S' + AS'^2)$$

$$\bar{J} = \varrho\, cJ + 2v'\bullet n(c^2\varrho' + \alpha\, S').$$

We assert that A can be chosen so large that both

J. Serrin

$$\varrho J \geqslant \text{Const.} \, (\varrho'^2 + v'^2 + S'^2) \qquad\qquad (\text{Const.} > 0)$$

and

$$\bar{J} \geqslant 0 \ .$$

Assuming for the moment that this can be done, the proof is completed by observing that then

$$\frac{d}{dt} \int \varrho \, J \leqslant \text{Const.} \int (\varrho'^2 + v'^2 + S'^2) \leqslant \text{Const.} \int \varrho \, J.$$

Integrating from $\ 0\ $ to $\ t\ $ yields

$$\int \varrho J \leqslant (\int \varrho \, J)_{t=0} \ \ e^{\text{Const.} \, t} = 0.$$

Thus since $\varrho J \geqslant 0$, it follows that $\varrho J = 0$. This in turn obviously implies $\varrho' = v' = S'$, and the two flows are identical.

The proof will thus be completed as soon as we verify the assertion. Since ϱ and $\ c\ $ are both continuous positive functions they have positive upper and lower bounds in $\ K\ $ during the time interval $\ 0\ $ to $\ t\ $. The first assertion is then clear, for the constant $\ A\ $ can certainly be chosen so that

$$\frac{c^2}{\varrho} \, \varrho'^2 + \frac{2\alpha}{\varrho} \, \varrho' S' + \varrho \, A S'^2 \geqslant \text{Const.} \, (\varrho'^2 + S'^2).$$

For the second assertion we observe that

$$\bar{J} \geqslant \varrho c J - 2 \, |v'| \cdot |c^2 \varrho' + \alpha S'| \ .$$

Now consider the two quadratic forms (in $\varrho', |v'|, S'$)

J. Serrin

$$\overline{J}_1 = \rho cJ + 2c^2 |v'| \, \rho' + 2\alpha |v'| S'$$

$$\overline{J}_2 = \rho cJ - 2c^2 |v'| \, \rho' - 2\alpha |v'| S' \ .$$

Certainly $\overline{J}$ equals either $\overline{J}_1$ or $\overline{J}_2$. If we can show that both $\overline{J}_1$ and $\overline{J}_2$ are non-negative for suitably large A , then $\overline{J} \geqslant 0$ and we are done. But by a direct calculation one finds that the eigenvalues of both $\overline{J}_1$ and $\overline{J}_2$ are

$$0, \ \rho c(1 + \frac{c^2}{\rho^2}) + 0 \, (\frac{1}{A}) \ , \ \rho cA + 0 \, (\frac{1}{A}) \ .$$

Hence if A is large enough all the eigenvalues will be non-negative. Thus $J_1 \geqslant 0$ and $J_2 \geqslant 0$ for large A , and the proof is complete.

J. Serrin

REFERENCES : CHAPTER IV

1] A.Signorini, Sopra alcune questioni di statica dei sistemi continui. Ann. Scuola Norm. Pisa 2 (1933), 231-257.

2] A.Signorini, Alcune proprietà di media nella elastostatica ordinaria. Rend. Lincei (6) 15 (1932), 151-156.

3] G.Grioli, Limitazioni per lo stato tensionale di un qualunque sistema continuo. Ann. di Mat. 39 (1955), 255-266.

4] C.Truesdell, Kinematics of Vorticity. Indiana University Press, 1954.

5] J.Serrin, Mathematical principles of classical fluid mechanics. Handbuch der Physik, Vol. 8/1. Springer, 1957. Especially $\S\S$ 17, 26, 28, 40, 72, 73, 74.

6] E.Foá, Sull'impiego dell'analisi dimesionale nello studio del moto turbolento. L'Industria (Milan) 43 (1929), p.426.

7] J.Serrin, On the stability of viscous fluid motion. Arch. Rat. Mech. Anal. 3 (1959), 1-13.

8] L.Payne and H.Weinberger, An exact stability bound for Navier-Stokes flow in a sphere. Nonlinear Problems, p.311-312, edited by R.E.Langer. Univ. of Winsconsin Press, 1963.

9] D.Graffi, Il teorema di unicità nella dinamica dei fluidi compressibili. Journ.Rat.Mech.Anal. 2 (1953), 99-106.

10] D.Graffi, Sul teorema di unicità nella dinamica dei fluidi, Annali di Mat. 50 (1960), pp.379-388.

11] D.Graffi, Sul teorema di unicità per le equazioni del moto dei fluidi compressibili in un dominio illimitato, Atti Acad.Sci. dell'Istituto Bologna, Sci-Fis. Classe, Series XI, 7 (1960), pp.1-8.

J.Serrin

[12] D.Graffi, Ancora sul teorema di unicità per le equazioni del moto dei fluidi, Atti Acad. Sci. dell'Istituto Bologna, Sci-Fis. Classe, Series XI, $\underline{8}$ (1961), pp. 7-14.

[13] D.Edmunds, On the uniqueness of viscous flows, Arch. Rat. Mech. Anal. (to appear 1963).

[14] J.Serrin, On the uniqueness of compressible fluid motions. Arch.Rat. Mech.Anal. $\underline{3}$ (1959), 271-288.

[15] K.O.Friedrichs, Symmetric Hyperbolic linear differential equations. Comm.Pure Appl. Math. $\underline{7}$ (1954), 345-392. Cf. also J.Leray, Hyperbolic differential equations. Lecture notes, Institute for Advanced Study, Princeton, 1952.

J. Serrin

V. THE INITIAL VALUE PROBLEM FOR THE NAVIER-STOKES EQUATION

As has been evident throughout this course, a selection of material frequently had to be made. In illustrating the comparison method it was necessary to restrict discussion to two or three main areas, with just passing mention being given to other applications of the method. Nevertheless, an attempt was made to include examples which ulluminated the more important ideas, while still staying near some physical problem of interest. In the same way, in discussing averaging methods a choice has to be made. Because of the remarkable amount of interest in the initial value problem for the Navier-Stokes equations, and because it illustrates very well the application of space averages to a difficult nonlinear problem, we have chosen this as the main object of study in the final lectures.

In what follows I shall confine myself for the most part to a somewhat restricted version of the problem. Consider a fixed (bounded or unbounded) domain Ω in E^n , where $n = 2$ or 3 . The domain Ω may be thought of as a rigid vessel filled with incompressible fluid; the fluid is initially set into motion, and (as in the preceding lectures) we are interested in the subsequent motion, subject to the Navier-Stokes equations and the condition of adherence at the boundary of Ω . More precisely, it is required to find a velocity field $v = v(x, t)$ and a pressure field $p = p(x, t)$ which for $x \in \Omega$ and for $t > 0$ satisfy the differential equations

$$v_t + v \quad \mathrm{grad}\ v = -\mathrm{grad}\ p + \Delta v$$

$$\mathrm{div}\ v = 0,$$

and obey in come sense the initial condition

J. Serrin

$$v(x, 0) = v_0(x) , \qquad\qquad x \in \Omega$$

and the boundary condition

$$v(x, t) = 0 , \qquad\qquad x \in \partial\Omega$$

In these equations we have set the external force $f = 0$ and the density and viscosity equal 1 for simplicity. Historically, it was C. W. Oseen and J. Leray who, some thirty years ago, first interested mathematicians in this problem; their work, moreover, still retains its importance. More recently, in 1951 Eberhard Hopf discovered that the problem could be studied using methods of functional analysis, and several years later Kiselev and Ladyzhenskaya published a further paper which set the modern trend in the subject. Many papers have appeared since then, extending the work of these authors.

We shall begin our discussion by considering the notion of <u>weak solution</u>, which is fundamental to all the modern work. Next we shall review the fundamental existence theorems of Hopf, and of Kiselev and Ladyzhenskaya. In the final lecture we shall consider the more delicate problems of regularity and uniqueness naturally associated with the concept of weak solution. In this work, which is in many ways rather abstract, I think you will see that the underlying idea is just that discussed in our opening lecture on averaging, namely, that it is valuable to consider the differential equation in an integrated form, and that the energy average is a natural tool to use in the theoretical investigation of a dynamical problem.

<u>A.</u> The program outlined above is fairly sophisticated, and requires a certain degree of preparation. We shall therefore begin with some definitions.

Let $u = u(x, t)$, $v = v(x, t)$ be vectors defined in $R = \Omega \times [0, \infty]$. We write

$$(u, v) = \int_{\Omega} u \cdot v \, dv$$

J. Serrin

for the spatial inner product of u and v . Clearly (u, v) is a function of
t ;when we wish to make the dependence on t more explicit we shall write
the product in the form (u(t), v(t)) . A vector v will be called <u>weakly divergence free</u> if

$$(v, \text{ grad } \psi) = 0 \qquad\qquad 0 \le t < \infty$$

for every function $\psi = \psi(x)$ which is continuously differentiable and vanishes near the boundary of Ω . Clearly if v is differentiable and satisfies
div v = 0 then it will satisfy the preceding condition; however, it is also clear
that a vector can satisfy the condition without having any derivatives in the ordinary sense.

Now let $\mathcal{D}(R)$ denote the family of all twice continuously differentiable vectors $\phi(x, t)$ in R such that $\text{div } \phi = 0$, and $\phi = 0$ near the boundary of Ω . A vector v = v(x, t) will be called a <u>weak solution of the initial
value problem</u> if it is weakly divergence free <u>and</u> if

$$\int_0^T \left\{ (v, \phi_t) + (v \cdot \text{grad } \phi, v) + (v, \Delta \phi) \right\} dt$$

$$= (v(T), \phi(T)) - (v_o, \phi_o) ,$$

for each T > 0 and each vector $\phi \in \mathcal{D}(R)$. Here $\phi_o = \phi(x, 0)$.

It is easy to see that this equation results from a direct averaging process on the Navier-Stokes equations, where $\phi(x, t)$ is the weighting function,
cf. Chapter IV, Part A[*]. Thus any solution of the Navier-Stokes equations

[*] In particular, one multiplies the Navier-Stokes equation by the vector ϕ ,
integrates over Ω ,and then finally integrates with respect to t from 0
to T .The integral form then results at once if we observe that, since ./.

J. Serrin

also satisfies this integral form of the equation. On the other hand, as already remarked previously when we discussed vectors which are weakly divergence free, it is clear that a vector may be a weak solution of the Navier-Stokes equations without being an ordinary solution. Finally, if v is a weak solution and if v has continuous derivatives, then v is an ordinary solution. To see this, we merely have to reverse the steps by which the integral equation was obtained, and use the well known device of the calculus of variations by which the Euler equation is obtained from the variational condition. The important point observed originally by Leray is that it is _easier_ to prove the existence of a weak solution then to prove the existence of an ordinary solution. Of course, the problem remains whether a weak solution can be considered as a genuine fluid motion, but at least the original problem is now reduced to two parts, each of which can be considered separately.

One final definition is necessary before we can state then major results of Hopf and of Kiselev and Ladyzhenskaya.

Definition. A vector $v = v(x, t)$ will be said to be in the class V if and only if for each $T, \ 0 < T < \infty$, it is in the closure of $\mathcal{D}(R)$ under the norm

$$\int_0^T (|\phi|^2 + |D\phi|^2) \, dt$$

$./. \quad \phi \in \mathcal{D}(R),$

$$(\text{grad } p, \phi) = 0, \quad (v \cdot \text{grad } v, \phi) = \cdot(v \cdot \text{grad } \phi, v), \quad (\Delta v, \phi) = (v, \Delta \phi),$$

and

$$\int_0^T (v_t, \phi) \, dt = - \int_0^T (v, \phi_t) \, dt + (v(T), \phi(T)) - (v_o, \phi_o).$$

J. Serrin

where

$$D\phi^2 = \int \mathrm{grad}\,\phi : \mathrm{grad}\,\phi\; dv \quad .$$

We observe that any vector in V has zero boundary data in the generalized sense, that is, it is the limit in norm of continuously differentiable functions which are zero on the boundary of Ω .. Moreover if $v \in V$ then v has a generalized gradient, denoted by grad v , which is defined to be the limit in the norm of the corresponding tensors grad ϕ . It can be shown that such generalized gradients obey the ordinary rules of calculus, though we shall not need this fact here. (The calculus of generalized derivatives is discoussed in many places; the reder may be referred specifically to references $[7]$ - $[8]$.)

THEOREM 1 (Hopf). For any weakly divergence free initial vector field of v_0 which vanishes near Ω there exists a weak solution $v \in V$ of the initial value problem. Moreover,

$$\frac{1}{2}\,|v|^2 + \int_0^t |Dv|^2\, dt \leq \frac{1}{2}\,|v_0|^2 \quad ,$$

that is, the sum of the kinetic energy and dissipated energy is less than or equal to the initial kinetic energy.

The proof is a beautiful application of the technique of Fourier approximation, unfortunately too long to include here. Nevertheless, we may observe that the process succeeds precisely because the <u>original</u> problem admits a formal energy identity

$$\frac{1}{2}\,|v|^2 + \int_0^t |Dv|^2\, dt = \frac{1}{2}\,|v_0|^2 \quad .$$

[A proof, assuming that v is a continuously differentiable solution, follows

J. Serrin

by integrating the energy transfer formula (8)

$$\frac{d}{dt} \int \frac{1}{2} q^2 \, dx = - \int T : D \, dv = - \int \mathrm{grad}\ v : \mathrm{grad}\ v \, dv,$$

that is

$$\frac{d}{dt} (\frac{1}{2} |v|^2) = - |Dv|^2 . \qquad \Big]$$

Two things should be noted. First, that is the <u>a priori</u> boundedness of the energy and stress <u>averages</u> which makes the proof work (it is exactly the quantities on the left side of the energy identity which are the building blocks of the space V). And second, that the rigorous proof given by Hopf yields only an energy inequality, not an energy identity.

The fact that the energy identity leads to a weak solution in the space V, leads us to expect that if we can obtain stronger <u>a priori estimates</u> for the norms (averages) of a solution, then we can correspondingly obtain the existence of a solution with more nearly classical behavior, that is, one which is not as "weak" as the solution found by Hopf. This is, in fact, exactly what Kiselev and Ladyzhenskaya did. Their result is as follows.

THEOREM 2. For any twice differentiable initial vector field v_0 which vanishes near Ω , there exists a weak solution $v \in V$ of the initial value problem, and a positive number T such that $|v_t|$ and $|Dv|$ are uniformly bounded in the interval $0 \leq t < T$. Moreover

$$\frac{1}{2} |v|^2 + \int_0^t |Dv|^2 \, dt = \frac{1}{2} |v_0|^2 .$$

The proof depends on Fourier approximation techniques, exactly as before. We can, however, present the <u>formal procedure</u> by which the necessa-

J. Serrin

ry <u>a priori</u> estimates are obtained. This is the important part of the proof[*].

By differentiating the Navier-Stokes equation with respect to t we get

$$v_{tt} + v_t \bullet \text{grad } v + v \bullet \text{grad } v_t = -\text{grad } p_t + \Delta v_t \ .$$

Next multiplying by v_t and integrating over Ω yields

$$(v_t, v_{tt}) + (v_t \bullet \text{grad } v, \ v_t) + (v \bullet \text{grad } v_t, v_t) = (\Delta v_t, v_t) \ ,$$

since $(\text{grad } p_t, \ v_t) = -(p_t, \ \text{div } v_t) = 0$. The previous equation can obviously be written

$$\frac{1}{2} \frac{d}{dt} |v_t|^2 = -(v_t \bullet \text{grad } v, \ v_t) - |Dv_t|^2 \ .$$

We next estimate the size of the first term on the right; thus

$$-(v_t \bullet \text{grad } v, \ v_t) \leq |v_t|_4^2 \ |Dv| \ ,$$

using Hölder's inequality, where $|f|_4 = (\int f^4 \ dx)^{1/4}$. Now according to a theorem of Sobolev the L_4 norm of a function can be estimated in terms of the norm of its first derivatives. In particular (cf. $[12]$)

$$|f|_4^2 \leq \begin{cases} \frac{1}{\sqrt{2}} |f| \cdot |Df| & n=2 \\[2ex] \frac{1}{3^{3/4}} |f|^{1/2} |Df|^{3/2} & n=3 \ . \end{cases}$$

---

[*] Indeed, as we have already indicated, it is just these a priori estimates which allow us to carry through the solution <u>process</u>, and which determine the <u>type</u> of weak solution which we obtain.

Hence

$$-(v_t \cdot \text{grad } v, v_t) \le \begin{cases} \dfrac{1}{\sqrt{2}} \, |v_t| \cdot |Dv_t| \cdot |Dv| \\[2ex] \dfrac{1}{3^{3/4}} \, |v_t|^{\frac{1}{2}} |Dv_t|^{\frac{3}{2}} |Dv| \end{cases}$$

$$\le \begin{cases} 2^{-3} \, |v_t|^2 |Dv|^2 + |Dv_t|^2 \\[2ex] 2^{-8} \, |v_t|^2 |Dv|^4 + |Dv_t|^2 \ , \end{cases}$$

Substituting into the earlier formula yields now

$$\frac{1}{2} \frac{d}{dt} |v_t|^2 \le \begin{cases} 2^{-3} \, |Dv|^2 |v_t|^2 & n = 2 \\[2ex] 2^{-8} \, |Dv|^4 |v_t|^2 & n = 3 \ . \end{cases}$$

This is the inequality which in the work of Kiselev and Ladyzhenskaya takes the places of the energy transfer formula in the paper of Hopf.

The next step is to integrate this differential inequality. In case $n=2$ the result is easily seen to be

$$|v_t| \le |v_{to}| \, \exp \left\{ 2^{-3} \int_0^t |Dv|^2 \, dt \right\} \le \text{Const.}$$

Since also

$$|Dv|^2 = -\frac{1}{2} \frac{d}{dt} |v|^2 \le |v| \cdot |v_t| \le \text{Const.}$$

the result for $n = 2'$ is completely proved. One even sees that we can take $T = \infty$, since the estimates are uniformly valid for all time.

J. Serrin

When $n = 3$, we have using the preceding inequality $|Dv|^2 \leq |v| \cdot |v_t|$,

$$\frac{d |v_t|}{dt} \leq 2^{-8} |v|^2 |v_t|^3 \leq 2^{-8} |v_o|^2 |v_t|^3 .$$

Dividing both sides by $|v_t|^3$ and integrating now gives

$$- \frac{1}{2} (|v_t|^{-2} - |v_{to}|^{-2}) \leq 2^{-8} |v_o|^2 \, t ,$$

that is

$$|v_t|^2 \leq \frac{|v_o|^2}{1 - 2^{-7} |v_o|^2 |v_{to}|^2 t} , \qquad (2^{-7} |v_o|^2 |v_{to}|^2 t < 1).$$

Thus $|v_t|$ is bounded as long as

$$t \leq 2^7 |v_o|^{-2} |v_{to}|^{-2} .$$

Since $|Dv|^2 \leq |v| \cdot |v_t|$, the required estimates are thus proved, with T being any number less than $2^7 |v_o|^{-2} |v_{to}|^{-2}$.

A rigorous proof of the theorem of Kiselev and Ladyzhenskaya remains a technical matter which can be handled by Hopf's method. The main idea of the proof, however, is the a priori estimate which we have just derived.

In conclusion, we whish to point out an important open problem. In two dimensions, the a priori estimate holds with $T = \infty$, and correspondingly, as we shall see in the final lecture, a genuine solution of the initial value problem exists for all $t > 0$. However, in three dimensions, the a priori estimates last only until some finite time T, and correspondingly a classical

J. Serrin

solution of the initial value problem is only known to exist in the interval $0 \leqslant t < T$. It is of the greatest physical and theoretical importance to know whether this breakdown of the solution process at the time T is simply due to an incomplete theory, or whether it is an inherent part of the problem. The answer will involve either a new existence theorem (ie. stronger a priori estimates), or else a counterexample to show that classical solutions need not persist for all time. Leray proposed the problem of finding such an example (cf. p. 225 of $[4]$), and in fact gave a possible method for its construction. Moreover, the problem has not yet been solved, and no one knows which way the answer will go.

B. The solutions obtained by Hopf and by Kiselev and Ladyzhenskaya are of course not necessarily solutions of the initial value problem in the sense originally intended. It is thus necessary to say a few words clarifying the relation between these solutions and a classical solution.

Let us assume to begin with that $v \in V$ is a weak solution of the initial value problem, that is, that

$$\int_0^T \left\{ (v, \phi_t) - (v \cdot \mathrm{grad}\ v, \phi) + (v, \Delta \phi) \right\} dt$$

$$= (v(T), \phi(T)) - (v_0, \phi_0),$$

For each $T > 0$ and each $\phi \in \mathcal{D}(R)$. Now corresponding to any function $u(x, t)$ let us introduce the space-time average

$$u_h = u_h(x, t) = \int K(\xi, \tau)\ u(x + \xi, t + \tau)\ d\xi\ d\tau$$

where the kernel $K(\xi, \tau)$ is a smooth non-negative function with the pro-

J.Serrin

perty that $K(\xi, \tau) \equiv 0$ outside a sphere of radius h about 0, while

$$\int K(\xi, \tau)\, d\xi\, d\tau = 1 \quad .$$

Thus u_h is a function whose values are <u>averages</u> of u over a sphere of radius h centered at (x, t). It is clear that the translation of a weak solution is again a weak solution, hence, in abbreviated notation,

$$\iint \left\{ v(x + \xi,\ t + \tau)\, \phi_t(x, t) + \ldots\ldots \right\}\, dx\, dt = \ldots \quad .$$

Multiplying both sides by $K(\xi, \tau)$, integrating with respect to ξ, τ, and finally reversing the orders of integration, then yieds

$$\int_0^T \left\{ (v_h, \phi_t) - ((v \cdot \mathrm{grad}\ v)_h, \phi) + (v_h, \Delta\phi) \right\}\, dt = 0$$

provided that ϕ vanishes when $t = 0$ and $t = T$.

Now since v_h is continuously differentiable with respect to both x and t, the preceding equation can be integrated by parts to give

$$\int_0^T \left\{ (v_{ht}, \phi) + ((v \cdot \mathrm{grad}\ v)_h, \phi) - (\Delta v_h, \phi) \right\}\, dt = 0 \quad .$$

Thus by standard techniques of the calculus of variations, since ϕ is an arbitrary function in $\mathcal{D}(R)$, we obtain the differential equation

$$v_{ht} + (v \cdot \mathrm{grad}\ v)_h = -\ \mathrm{grad}\ p_h + \Delta v_h \quad .$$

J. Serrin

This shows that the <u>averaged</u> velocity $\mathbf{V}_h$ is approximately a solution of the Navier-Stokes equations, or looked at in another way, the velocity v satisfies the <u>Reynolds average form</u> of the Navier-Stokes equation.

If we form the curl of the preceding equation, and set $\omega = \text{curl } v$, we obtain the averaged vorticity equation

$$\omega_{ht} - \Delta \omega_h = \text{div}\,(\omega v - v \omega)_h \quad ,$$

where the right hand side is defined by its components $(\omega_i v_j - \omega_j v_i)_{h,i}$. Now let $k(x,t)$ denote the fundamental solution of the heat equation. Then clearly we have the following integral representation

$$\omega_h(x,t) = \int \text{grad } k(x - \xi, t - \tau) \quad (\omega v - v\omega)_h \; d\xi \, d\tau + B_h(x,t) ,$$

where $B_h(x,t)$ is a solution of the heat equation. Finally letting $h \to 0$ yields the formula

$$(\ast) \qquad \omega(x,t) = \int \text{grad } k(x - \xi, t - \tau) \cdot (\omega v - v\omega) \; d\xi \, d\tau + B(x,t) ,$$

where $B(x,t)$ is again a solution of the heat equation. We can now state the fundamental regularity theorem for weak solutions of the Navier-Stokes equation [9] .

THEOREM. The weak solution of Kiselev and Ladyzhenskaya is continuously differentiable in the space variables, and Lipschitz continuous in time, and satisfies the Navier-Stokes equation almost everywhere in $R = \Omega \times (0, T)$.

The proof is too long to include here, but the main idea is to consider $(\ast)$ as a <u>linear integral equation</u> for ω , the function v being considered fixed. For the details of the proof the reader is referred to reference [8] .

128

J. Serrin

It may be added that the same process fails for the Hopf solution because in this case the kernel of the integral equation (which involves v) is not sufficiently regular. Thus the additional properties of the Kiselev-Ladyzhenskaya solution are seen to be of crucial importance in finally establishing the existence of a differentiable solution of the Navier-Stokes equation.

If the boundary of Ω is smooth then stronger conclusions can be obtained. In particular, Ito has shown by quite different methods, the existence of a classical solution continuously taking on the given boundary values. However, his proof arc extremely complicated and one would like to obtain his results by means of the relatively simpler methods outlined here.

To conclude the course, it may be of interest to review some of the open problems which we have noticed.

1) To extend the comparison method in free boundary theory to non-symmetric flows

2) To exploit more fully the variational approach to the stability of la minar fluid motions

3) To obtain stronger uniqueness theorems for the initial value problem in exterior domains.

4) To obtain the existence of a suitably regular solution of the initial value problem for the Navier-Stokes equation in 3 dimensions which persists for all $t > 0$.

5) It would finally be worth while to be able to prove the existence theorems of Hopf and of Kiselev and Ladyzhenskaya using fixed point methods (cf. $\begin{bmatrix} 10 \end{bmatrix}$) rather than Fourier approximation.

There are of course countless other problems still open in the application of comparison and averaging methods, and it is hoped that some of you at least will find this a fruitful field of study .

J.Serrin

REFERENCES : CHAPTER V

[1] C.Oseen, Neuere Methoden und Ergebnisse der Hydrodynamik, Leipzig, 1927

[2] J.Leray, Étude de diverses equations intégrales non-linéares et de quelques problèmes que pose l'Hydrodynamique, Journ. Math. Pures Appl. 12 (1933), pp. 1-82.

[3] J.Leray, Essai sur les mouvements plans d'un liquide visqueux que limitent des parois, Journ. Math. Pures Appl. 13 (1934), pp.331-- 418.

[4] J.Leray, Sur le mouvement d'un liquide visqueux emplissant l'espace, Acta Math. 63 (1934), pp. 193-248.

[5] E.Hopf; Über die Anfangswertaufgabe fur die hydrodynamischen Grundgleichungen, Math. Nachrichten 4 (1951), pp.213-231.

[6] A.A.Kiselev and O.A.Ladyzhenskaya, On existence and uniqueness of the solution of the nonstationary problem for a viscous incompressible fluid, Izvestiya Akad. Nauk SSSR 21 (1957), pp. 655-680.

[7] C.B.Morrey, Multiple integral problems in the calculus of variations, Ann. di Pisa 14 , (1960).

[8] L.Nirenberg, Remarks on strongly elliptic partial differential equations. Comm. Pure Appl. Math. 8 (1955).

[9] J.Serrin, On the interior regularity of weak solutions of the Navier--Stokes equations, Arch. Rational Mech. Analysis 9 (1962), pp.187--195.

J. Serrin

[10] D. Gilbarg, Boundary value problems for nonlinear elliptic equations in n variables. Nonlinear Problems, edited by R. E. Langer. Univ. of Wisconsin Press, 1963.

[11] S. Ito, The existence and uniqueness of regular solution of nonstationary Navier-Stokes equation. Journ. Fac. Science, Univ. of Tokyo, $\underline{9}$ (1961), pp. 103-140.

[12] J. Serrin, The initial value problem for the Navier-Stokes equations. Nonlinear problems, edited by R. E. Langer. Univ. of Wisconsin Press, 1963. (This paper contains a large bibliography.)

CENTRO INTERNAZIONALE MATEMATICO ESTIVO
(C. I. M. E.)

H. ZIEGLER

THERMODYNAMIC ASPECTS OF CONTINUUM MECHANICS

ROMA - Istituto Matematico dell'Università

THERMODYNAMIC ASPECTS OF CONTINUUM MECHANICS
by
HANS ZIEGLER

1. <u>Classical thermodynamics</u>. Recent developments have made it clear that continuum mechanics cannot be separated from thermodynamics. In the second half of the last century statistical mechanics has been created in order to provide a mechanical basis for thermodynamic phenomena. Today the process is being reversed: we are turning to thermodynamics and statistical mechanics for the explanation of certain aspects of continuum mechanics. The borderline is inevitably reached in any attempt to give a complete outline of the fundamental laws of continuum mechanics.

In this first section we shall formulate the laws of classical thermodynamics in a manner fit for use in connection with mechanical problems [1] .

Let us consider a system the state of which is completely described by the mechanical coordinates x_k $(k = 1, 2, \ldots, n)$ and the temperature $\theta > 0$. (The state of an infinitesimal element of an elastic or a rigid/perfectly plastic body, e.g., is completely described by the strain components ε_{kl} and θ). The x_k and θ are the independent state variables; any function of them will be called a state function. If the work done on the system is given by

$$dW = X_k \, dx_k \quad , \quad \text{(1)} \tag{1.1}$$

the X_k are the forces corresponding to the mechanical state variables x_k. (For a volume element under infinitesimal strains the forces corresponding to the ε_{kl} are the stress components σ_{kl}).

<u>The first fundamental theorem</u> states that there exists a state function $\mathcal{U}\,(x_k, \theta)$, called the intrinsic energy of the system, such that

---

(1) We shall use the summation convention.

H. Ziegler

$$(1.2) \qquad dU = dW + dQ = X_k \, dx_k + dQ \quad ,$$

where dQ is the influx of heat.

The <u>second fundamental theorem</u> states that there exists another state function $S(x_k, \theta)$, called the entropy of the system, such that

$$(1.3) \qquad \theta \, dS \gtrless dQ \quad .$$

If (1.3) holds with the equality sign, the process is referred to as reversible, otherwise as irreversible. The theorem can also be stated in the form

$$(1.4) \qquad dS = d^{(r)}S + d^{(i)}S$$

due to Carnot and Clausius, where

$$(1.5) \qquad d^{(r)}S = \frac{dQ}{\theta}$$

is called the influx of entropy and

$$(1.6) \qquad d^{(i)}S > 0$$

the entropy production inside the system, zero for reversible processes and positive for irreversible ones. The last statement justifies the use of the superscripts r and i for the reversible and irreversible parts of the process.

From (1.2), (1.5) and (1.4) we deduce

$$(1.7) \qquad dW = dU - dQ = dU - \theta d^{(r)}S = dU - \theta dS + \theta d^{(i)}S \quad .$$

H. Ziegler

On account of (1.1) and the fact that U and S are state functions, (1.7) is equivalent with the relation

$$(1.8) \qquad X_k\, dx_k = \left(\frac{\partial U}{\partial x_k} - \theta\,\frac{\partial S}{\partial x_k}\right) dx_k + \left(\frac{\partial U}{\partial \theta} - \theta\,\frac{\partial S}{\partial \theta}\right) d\theta + \theta d^{(i)}S \quad,$$

which thus is a direct consequence of the fundamental theorems and hence is valid for any process. For pure heating or cooling $dx_k = 0$. In this case (1.8) reduces to

$$(1.9) \qquad \left(\frac{\partial U}{\partial \theta} - \theta\,\frac{\partial S}{\partial \theta}\right) d\theta + \theta d^{(i)}S = 0 \quad.$$

On account of (1.6) the second therm is non-negative, while the quantity between brackets is a state function and hence is independent of $d\theta$. Since (1.9) must hold for positive and negative values of $d\theta$, it follows that, independent of the type of process,

$$(1.10) \qquad \frac{\partial U}{\partial \theta} - \theta\,\frac{\partial S}{\partial \theta} = 0$$

and that, for the process considered here, $d^{(i)}S = 0$, i.e., that heating and cooling are reversible phenomena. The differential equation (1.10) establishes a connection between the intrinsic energy and the entropy of the system. Making use of (1.10) and of the notations

$$(1.11) \qquad \frac{\partial U}{\partial x_k} - \theta\,\frac{\partial S}{\partial x_k} = X_k^{(r)}$$

H. Ziegler

and

$$(1.12) \qquad X_k - X_k^{(r)} = X_k^{(i)} \quad ,$$

we obtain, instead of (1.8),

$$(1.13) \qquad \theta d^{(i)}S = X_k^{(i)} dx_k \gtreqless 0 \quad ,$$

where the statement concerning the sign follows from (1.6). Inserting (1.5) and (1.13) in (1.4), we also have

$$(1.14) \qquad dS = \frac{1}{\theta} X_k^{(i)} dx_k + \frac{dQ}{\theta} \quad .$$

According to (1.12) each force X_k appears as the sum of two terms $X_k^{(r)}$ and $X_k^{(i)}$. On account of (1.13) the entropy production inside the system is completely determined by the $X_k^{(i)}$. It is therefore reasonable to refer to the $X_k^{(i)}$ as the irreversible forces and to the $X_k^{(r)}$ as the reversible ones. If, as an additional state function, we introduce the free energy

$$(1.15) \qquad F = U - \theta S,$$

we obtain from (1.11) and (1.10)

$$(1.16) \qquad \frac{\partial F}{\partial x_k} = X_k^{(r)} \quad , \qquad \frac{\partial F}{\partial \theta} = -S \ .$$

Apart from its sign, the free energy thus serves as a potential function for the reversible forces and the negative entropy. It follows that the reversible forces are state functions.

Sometimes a process is conducted in such a way that θ is a given

H. Ziegler

function of the x_k . (In an isothermal process θ = const , in an isentropic process S = const). In such cases $-\bar{F}$ has the properties of a mechanical potential.

It follows from (1.13) that any $X_k^{(i)}$ changes sign whenever the corresponding dx_k is reversed. In consequence the $X_k^{(i)}$ are not state functions, but depend on the velocities $\dot{x}_k$. Besides, they may depend on the state of the system and on its history.

Classical thermodynamics does not provide any clue as to the dependence $X_k^{(i)}(\dot{x}_j)$. For linear relationships between the velocities and the irreversible forces,

$$(1.17) \qquad X_k^{(i)} = a_{kj}\,\dot{x}_j \quad ,$$

Onsager $\begin{bmatrix}2\end{bmatrix}$ has established the symmetry relations

$$(1.18) \qquad a_{jk} = a_{kj} \quad .$$

They are based on a statistical treatment of systems moving freely in the vicinity of an equilibrium configuration. Onsager's demonstration makes use of a principle of microscopic reversibility and of some additional assumptions.

In continuum mechanics many processes are irreversible (particularly on account of interior friction). However, we are usually not concerned with infinitesimal free motions about an equilibrium position, but rather with finite processes taking place under given forces (e.g., with the deformation of an element of a plastic body under prescribed stresses). In the more interesting cases the relationship between velocities and irreversible forces is not linear. Sometimes (e.g., in a plastic body) it is even impossible to linearize it. There exists thus a definite need for a generalization of Onsager's theory.

H. Ziegler

In fact, classical thermodynamics is little more than a theory of thermostatic equilibrium, restricted to certain special cases, and a really dynamic
theory does not exist in this field. This has been emphasizend by Truesdell
[3] in the following words : "It is suggested that an attempt be made to crea-
te and organize the logical structure of a true thermodynamics of irreversible
processes along the lines successfully employed two hundred years ago by Euler
and others in converting the unorganized special methods and principles of
seventeenth-century mechanics into the general theory we know today."

2. <u>Additional principles.</u> In this section a possibility of realizing Truesdell's program will be described. It consists in a generalization of Onsager's
principle $[1, 4, 5, 6]$, limited to processes which are slow compared with
the motions of the molecules involved.

In a system of the type considered in Section 1 the rate of entropy production $d^{(i)}S/dt$ depends on the velocities $\dot{x}_k$, on the state of the system
and possibly also on its history. On account of (1.13) the rate of dissipation
work is

$$(2.1) \qquad P^{(i)} = X_k^{(i)}\dot{x}_k = \theta \frac{d^{(i)}S}{dt} \gtreqless 0 \quad .$$

In a given state of the system, preceded by a given history, $P^{(i)}$ is a function of the velocities $\dot{x}_k$ alone and hence can be written

$$(2.2) \qquad P^{(i)} = D(\dot{x}_k) \gtreqless 0 \quad .$$

The function $D(\dot{x}_k)$ is referred to as the dissipation function of the system.
It must be considered as the primary quantity in an irreversible process and

H. Ziegler

is of similar importance for the irreversible part of the process as the state functions F or U are for its reversible part. The irreversible forces $X_k^{(i)}$ are secondary quantities, connected with D by the relation

$$(2.3) \qquad X_k^{(i)} \dot{x}_k = D(\dot{x}_k) \; ,$$

following from (2.1) and (2.2).

Let us interpret $D(\dot{x}_k)$ as a function in a euclidean velocity space $\dot{x}_k$ (Fig. 1) of n dimensions, and let us assume, for convenience, that $D(\dot{x}_k)$ be sufficiently regular. (For a more exact treatment see $[6]$). The dissipation function may be visualized by means of (hyper-)surfaces $D(\dot{x}_k)=M$, where M belongs to a set of non-negative constants. Furthermore, the $\dot{x}_k$ and the $X_k^{(i)}$ define two vectors in velocity space.

What we are looking for is a connection between the vectors $\dot{x}_k$ and $X_k^{(i)}$. From the first fundamental theorem we obtain no statement of use for this purpose. The second one yields the inequality (2.1) , implying that the scalar product of the two vectors is non-negative. In order to establish a more definite connection, let us stipulate the following

<u>Principle of least irreversible force</u> : If the value $M > 0$ of the dissipation function $D(\dot{x}_k)$ and the direction of the irreversible force $X_k^{(i)}$ are prescribed, the actual velocity $\dot{x}_k$ minimizes the magnitude of $X_k^{(i)}$ subject to the auxiliary condition (2.3).

In other words : Among all vectors $\dot{x}_k$ with end points P on a given D-surface, the projection of the real one (or ones) in the direction of $X_k^{(i)}$ is a maximum.

It follows immediately that $X_k^{(i)}$ is normal to the D-surface at P . A great deal of additional implications can be derived from the principle of

141

H. Ziegler

least irreversible force. In the remainder of this section some of them will be discussed without proof. (For the proofs see $\begin{bmatrix} 6 \end{bmatrix}$).

Provided the principle is valid, the surfaces $D(\dot{x}_k) = M$ are convex. Each one of them contains those with smaller values of M and hence also the origin. It follows that the vector $X_k^{(i)}$ has the direction of the exterior normal at P . The function D increases monotonically on any radius from 0 .

If the increase of D on any radius from 0 is sufficient, the projection of $X_k^{(i)}$ on the radius also increases. Let us restrict ourselves to systems subject to this condition and let us denote them as stable, for it can be shown that, whenever the condition is violated, self-sustained oscillations are apt to develop. In a stable system the last principle is equivalent with the following.

<u>Principle of maximum rate of dissipation work</u> : If the irreversible force $X_k^{(i)} \neq 0$ is prescribed, the actual velocity $\dot{x}_k$, subject to the auxiliary condition (2.3), maximizes the rate of dissipation work.

On account of (2.1) this principle can also be formulated as a <u>principle of maximum rate of entropy production.</u> In this form it appears as a natural and physically particularly plausible extension of the second fundamental theorem.

Another consequence of the principle of least irreversible force is the inequality

$$(2.4) \qquad X_k^{(i)} (\dot{x}_k - \dot{x}_k^{*}) \geqq 0,$$

valid for the actual irreversible force $X_k^{(i)}$, the actual velocity $\dot{x}_k$ and any other velocity $\dot{x}_k^{*}$ with $D(\dot{x}_k^{*}) \leqq D(\dot{x}_k)$. Still another representation of the connection between $\dot{x}_k$ and $X_k^{(i)}$, based on the assumed regularity

H. Ziegler

of the function $D(\dot{x}_k)$, is

$$(2.5) \qquad X_k^{(i)} = \left(\frac{\partial D}{\partial \dot{x}_j} \dot{x}_j\right)^{-1} D \frac{\partial D}{\partial \dot{x}_k} \quad .$$

Once the relation between $\dot{x}_k$ and $X_k^{(i)}$ is established, the dissipation function can be expressed, by

$$(2.6) \qquad D(\dot{x}_k) = D'(X_k^{(i)}) \; ,$$

in terms of the irreversible forces. It can then be shown that each one of the results stated above has a corollary, obtained by interchanging the roles of $\dot{x}_k$ and $X_k^{(i)}$. Thus the three principles can be reformulated, the first one as a <u>principle of least velocity</u>. The inequality corresponding to (2.4) is

$$(2.7) \qquad (X_k^{(i)} - X_k^{(i)*}) \dot{x}_k \geqq 0 \quad .$$

It holds for the actual velocity $\dot{x}_k$, the actual irreversible force $X_k^{(i)}$ and any other irreversible force $X_k^{(i)*}$ with $D'(X_k^{(i)*}) \leqq D'(X_k^{(i)})$. Finally the corollary of (2.5) reads

$$(2.8) \qquad \dot{x}_k = \left(\frac{\partial D'}{\partial x_j^{(i)}} X_j^{(i)}\right)^{-1} D' \frac{\partial D'}{\partial x_k^{(i)}}$$

If $D(\dot{x}_k)$ satisfies the functional equation

H. Ziegler

$$(2.9) \qquad \frac{\partial D}{\partial \dot{x}_k} \, \dot{x}_k = f(D) \quad ,$$

where $f(D)$ is arbitrary, the D-surfaces are similar and similarly situated with respect to the origin in velocity space. Let us refer to a dissipation function of this type as quasi-homogeneous. In this case (2.5) reduces to

$$(2.10) \qquad X_k^{(i)} = \frac{D}{f(D)} \, \frac{\partial D}{\partial \dot{x}_k} \quad ,$$

and this is equivalent with

$$(2.11) \qquad X_k^{(i)} = \frac{\partial \varphi}{\partial \dot{x}_k} \quad , \qquad \text{where} \qquad \varphi = \int \frac{D \, dD}{f(D)}$$

With $f(D) = r D$, (2.9) takes the form

$$(2.12) \qquad \frac{\partial D}{\partial \dot{x}_k} \, \dot{x}_k = r D \quad .$$

A function D satisfying (2.12) is called homogeneous of degree r . Here (2.10) yields

$$(2.13) \qquad X_k^{(i)} = \frac{1}{r} \, \frac{\partial D}{\partial \dot{x}_k} \quad .$$

In the particular case $r = 2$ the dissipation function is given by the quadratic form

H. Ziegler

$$(2.14) \qquad D(\dot{x}_k) = a_{jk}\, \dot{x}_j\, \dot{x}_k \quad ,$$

the generality of which is not restricted by setting

$$(2.15) \qquad a_{jk} = a_{kj}$$

On account of (2.13)

$$(2.16) \qquad X_k^{(i)} = a_{kj}\, \dot{x}_j$$

Thus Onsager's relations (1.17), (1.18) are obtained as a special case of the present theory.

3. <u>Thermodynamics and continuum mechanics</u>. It has been pointed out in Section 1 that it is impossible to separate continuum mechanics from thermodynamics. The reason fro this is the fact that, in continuum mechanics, the microstructure of the material under consideration remains indefined. In consequence it is impossible to formulate the work of the interior forces, entering the energy theorem of mechanics . This makes it necessary to replace this theorem by the first fundamental law, and thus thermodynamics is brought in even in cases where heat effects are negligible.

In this section we shall formulate the basic mechanical and thermodynamic equations for a continuum, using cartesian tensor notation.

Let y_j denote the cartesian coordinates and t the time. Let partial derivatives with respect to y_j or t be indicated by the subscript j or 0 respectively, preceded by a comma, and let the material derivative be denoted by a dot.

If ρ represents the density and v_j the velocity field, the <u>principle</u>

H. Ziegler

<u>of conservation of mass</u> for an arbitrary volume V requires that

$$(3.1) \qquad \int (\dot{\varrho} + \varrho \, v_{j,j}) dV = 0 \; .$$

(For a detailed derivation of this result and the next ones up to (3.9) see [7]). The differential form of (3.1), i.e., the principle of conservation of mass for an element, is

$$(3.2) \qquad \dot{\varrho} + \varrho \, v_{j,j} = \varrho_{,0} + (\varrho \, v_j)_j = 0 \; .$$

The <u>momentum theorem</u> for the volume V is given by

$$(3.3) \qquad \int \varrho \, \dot{v}_k \, dV = \int \varrho \, f_k \, dV + \int \sigma_{kl} \, \nu_l \, dS = \int (\varrho \, f_k + \sigma_{kl,l}) \, dV \quad ,$$

where S is the surface of V, ν_l its exterior unit normal, f_k the specific body force (i.e., the body force per unit mass) and σ_{kl} the stress tensor. The differential form of (3.3), i.e., the momentum theorem for a single element, reads

$$(3.4) \qquad \varrho \, \dot{v}_k = \varrho f_k + \sigma_{kl,l} \, .$$

The <u>angular momentum theorem</u> is similar and, in its differential form, establishes the symmetry of the stress tensor.

In order to replace the energy theorem of mechanics by the first fundamental theorem of thermodynamics, we note that the intrinsic energy contained in the volume V is

H. Ziegler

$$(3.5) \qquad \mathcal{U} = \int \varrho\, \mathcal{u}\, dV \;,$$

where $\mathcal{u}$ denotes the specific intrinsic energy, dependent on the mechanical state of the element, i.e., on its deformation, and on the temperature. The influx of heat into the volume V is

$$(3.6) \qquad \mathcal{Z}^{(\mathcal{u})} = -\int q_k\, \nu_k\, dS \;,$$

where the vector q_k denotes the heat flux. Starting from (1.2) and observing that, in a continuum, the energy of an element is composed of its kinetic and intrinsic energies, we state the <u>first fundamental theorem</u> for the volume V in the following form :

The material rate of increase of the sum of the kinetic and intrinsic energies in equal to the rate of work of the exterior forces plus the heat influx.

The analytical form of this statement is

$$\int \varrho(v_k\dot{v}_k + \dot{\mathcal{u}})dV = \int \varrho\, f_k v_k\, dV + \int (\sigma_{kl} v_k - q_l)\,\nu_l\; dS$$

$$(3.7) \qquad = \int (\varrho\, f_k v_k + \sigma_{kl} v_{k,l} + \sigma_{kl,l} v_k - q_{l,l})dV$$

On account of (3.4) and the symmetry of σ_{kl} (3.7) reduces to

$$(3.8) \qquad \int \varrho\, \dot{\mathcal{u}}\, dV = \int (\sigma_{kl} v_{kl} - q_{k,k})\, dV \;,$$

where

$$(3.9) \qquad v_{kl} = \frac{1}{2}(v_{k,l} + v_{l,k})$$

H. Ziegler

is the rate of deformation. Thus the differential form of the theorem, i.e., the first fundamental law for the element, reads

$$(3.10) \qquad \rho \dot{u} = \sigma_{kl} \, v_{kl} - q_{k,k} \quad .$$

In this form the analogy with (1.2) is complete.

Once the first fundamental theorem has been accepted, it is a matter of consequence to proceed to the second one. In order to do so, we introduce the entropy contained in the volume V ,

$$(3.11) \qquad S = \int \rho \, s \, dV \quad ,$$

where s denotes the specific entropy. Let us restrict ourselves to the case where s depends on the same variables as u , i.e., on the deformation and the temperature of the element. Comparing (3.6) with (1.5) , it is easy to see that the influx of entropy into the volume V is

$$(3.12) \qquad \mathcal{L}^{(s)} = - \int \frac{q_k}{\theta} \, v_k \, dS \, .$$

Combining (1.3) and (1.5) , we state the <u>second fundamental theorem</u> for the finite volume as follows :

The material rate of increase of entropy is equal to or greater than the entropy influx.

In other words : The entropy production inside V is non-negative. The analytical form of this statement is

$$(3.13) \qquad \dot{S} \gtreqless \mathcal{L}^{(s)}$$

H. Ziegler

or

$$(3.14) \qquad \int \rho \, \dot{s} \, dV \gtrless - \int \frac{q_k}{\theta} \, \nu_k \, dS = - \int (\frac{q_k}{\theta})_{,k} \, dV \ .$$

If (3.14) holds with the equality sign, the process is reversible, otherwise it is irreversible. In the last case entropy is produced inside V . The differential form of (3.14) is

$$(3.15) \qquad \rho \, \dot{s} \gtrless - (\frac{q_k}{\theta})_{,k} = \frac{q_k}{\theta^2} \, \theta_{,k} - \frac{q_{k,k}}{\theta} \quad .$$

In order to interpret this inequality, let us transfer the results of Section 1 to a single element of the continuum considered here, restricting ourselves, for convenience, to infinitesimal deformations. (For finite deformations see [6]). Here the infinitesimal strain components ε_{kl} are the mechanical state variables, and the stress components σ_{kl} are the corresponding forces for the unit volume. According to (1.12) the stress tensor can be represented, by

$$(3.16) \qquad \sigma_{kl} = \sigma_{kl}^{(r)} + \sigma_{kl}^{(i)} \quad ,$$

as the sum of a reversible and an irreversible part.

With $u(\varepsilon_{kl}, \theta)$ and $s(\varepsilon_{kl}, \theta)$ the reversible stress tensor $\sigma_{kl}^{(r)}$ is a state function. The relations (1.11) and (1.10) take the form

$$(3.17) \qquad \sigma_{kl}^{(r)} = \rho \, (\frac{\partial u}{\partial \varepsilon_{kl}} - \theta \, \frac{\partial s}{\partial \varepsilon_{kl}}) \ , \qquad \frac{\partial u}{\partial \theta} - \theta \, \frac{\partial s}{\partial \theta} = 0 \ .$$

H. Ziegler

Instead of (1.16) we now have

$$(3.18) \qquad \sigma_{kl}^{(r)} = \rho \, \frac{\partial f}{\partial \varepsilon_{kl}} \quad , \qquad -s = \frac{\partial f}{\partial \theta} \quad ,$$

where

$$(3.19) \qquad f = u - \theta s$$

is the specific free energy.

The irreversible stress tensor $\sigma_{kl}^{(i)}$ depends on the rate of deformation and possibly also on the state of the element and on its history. Any component of $\sigma_{kl}^{(i)}$ changes sign together with the corresponding component of V_{kl}. A comparison with (1.2) and (3.10) shows that the relation (1.14) becomes

$$(3.20) \qquad \rho \, \dot{s} = \frac{1}{\theta} \, \sigma_{kl}^{(i)} \, V_{kl} - \frac{q_{k,k}}{\theta} \quad .$$

Here the first term on the right-hand side represents the entropy production inside the element, due to the work of the irreversible stress tensor, while the second one describes the entropy influx, due to heat exchange with the environment. The sum $\sigma_{kl}^{(i)} \, V_{kl}$ is the rate of dissipation work and indicates the rate at which the work done on the element is transformed into heat.

Writing (3.20) in the form

$$(3.21) \qquad \rho \, \dot{s} = \frac{1}{\theta} \, \sigma_{kl}^{(i)} \, V_{kl} - (\frac{q_k}{\theta})_{,k} - \frac{q_k}{\theta^2} \, \theta_{,k}$$

and integrating over V , we obtain

H. Ziegler

$$(3.22) \qquad \dot{S} = \int \rho \, \dot{s} \, dV = \int \frac{1}{\theta} \, \sigma_{kl}^{(i)} \, V_{kl} \, dV - \int \frac{q_k}{\theta^2} \, \theta_{,k} \, dV - \int \frac{q_k}{\theta} \, \nu_k \, dS.$$

On account of (3.12) the last term on the right-hand side represents the entropy influx, due to heat exchange with the environment. It follows that the two other terms describe the entropy production inside V . The first one obviously represents the entropy production due to the work of the irreversible stress tensor, the second one the entropy production due to heat exchange inside V .

It is interesting to compare (3.20) and (3.22) . In (3.20) the term representing intrinsic heat exchange is not present, and this is clearly due to the fact that the element is characterized by a single temperature. Thus any kind of heat exchange inside the finite volume V is indeed an irreversible process, accompanied by an entropy production. For the single element, however, the same process appears as reversible , since no entropy is produced in its interior. The apparent paradox is easily solved by considering the boundaries between the elements as the sources of entropy production due to heat exchange. It follows, however, that statements concerning entropy production must be handled with caution : any such statement, although valid for the single elements, need not necessarily hold for a finite volume, and vice versa.

· With (3.21) the inequality (3.15) reduces to

$$(3.23) \qquad \sigma_{kl}^{(i)} \, V_{kl} - \frac{q_k}{\theta} \, \theta_{,k} \geq 0 \qquad .$$

The left hand side is the rate of entropy production per unit volume. It consists of the entropy production within the element and the element's share

H. Ziegler

of the entropy production in the boundaries. On account of the presence of the second term, (3.23) cannot be considered as the expression of the second fundamental theorem for the single element, although it is the differential form of this theorem for the finite volume. However, since the two terms in (3.23) represent entropy productions of entirely different sources, it is to be expected that they are independent of each other and that, in consequence, each one of them must be non-negative. In fact, it is clear that the irreversible stress tensor $\sigma_{kl}^{(i)}$, as a function of the deformation rate V_{kl} , the state of the element and possibly its history, is independent of the surrounding elements and hence of the temperature gradient $\theta,_{k}$. It is equally plausible that the heat flux q_k depends solely on the differences between the states of adjacent elements but not on the instantaneous deformation rate of a single element. It follows that (3.23) must be split up into

$$(3.24) \qquad \sigma_{kl}^{(i)} V_{kl} \geqq 0 \qquad \text{and} \qquad - q_k \, \theta,_k > 0 \ .$$

The first inequality represents the second fundamental theorem for the element. It states that the entropy production within the element, due to the work of the irreversible stress tensor, is non-negative. The second inequality may be considered as the expression of the same theorem for the boundaries between the elements. It states that any entropy production due to heat exchange is non-negative.

4. <u>Constitutive equations</u>. The basic equations formulated in the last section are valid for arbitrary continua. For any specific material they must be supplemented by the proper constitutive equations, connecting the kinematic variables (such as strain, rate of deformation, etc.) with the static ones

H. Ziegler

(stress, stress rate, etc.). It is clear that these constitutive equations must be consistent with the general laws, in particular with the fundamental theorems of thermodynamics. In this section we shall discuss some implications of this postulate.

In _elasticity_ some authors (see, e.g., [8]) distinguish between three different types of material. Although the definitions are usually given in terms of finite deformations, it seems possible without loss of any essential feature to discuss them in terms of infinitesimal strains ε_{kl} and the temperature θ as state variables. In order to get rid of the temperature and of the necessity to take heat exchange into consideration, one usually assumes that the process is conducted in such a manner that θ is either constant (isothermal process) or a given function of the strain history (as, e.g., in an adiabatic process). For an anisotropic material the definitions then are essentially the following ones :

The hypoelastic body is defined by a linear relation,

$$(4.1) \qquad d\sigma_{ij} = C_{ijkl}(\sigma_{mn}) \, d\varepsilon_{kl} \quad ,$$

between the increments of strain and stress. The elastic body is defined by a relation

$$(4.2) \qquad \sigma_{ij} = g_{ij}(\varepsilon_{kl})$$

between strain and stress. If this relation has the form

$$(4.3) \qquad \sigma_{ij} = \rho \, \frac{\partial f(\varepsilon_{kl})}{\partial \varepsilon_{ij}} \quad ,$$

H. Ziegler

where   - f   denotes the specific potential energy, the body is called hyperelastic.

It is evident that, with these definitions, any hyperelastic body is elastic,
and that any elstic body is hypoelastic. It is usually maintained that the reverse is not true, and from a purely mathematical point of view this statement
is clearly correct. By simple thermodynamic reasoning, however, it is easy
to see that any hypoelastic body is elastic, and that any elastic body is hyperelastic, so that there is no point in distinguishing between the three types of
material.

From the viewpoint of thermodynamics it is reasonable to retain θ
as an indipendent state variable and to generalize the definitions (4.1) through
(4.3) accordingly. Let the hypoelastic body be defined by the genralization

$$(4.4) \qquad d\sigma_{ij} = C_{ijkl}(\sigma_{mn}, \theta)\, d\varepsilon_{kl} + G_{ij}(\sigma_{mn}, \theta)\, d\theta$$

of (4.1) . If the sign of $d\varepsilon_{kl}$ is changed, this affects $d\sigma_{ij}$ but does
not reverse the sign of any finite part of σ_{ij} . It follows from Section 3
that the stress tensor is reversible, i.e. that

$$(4.5) \qquad \overset{(r)}{\sigma}_{ij} = \sigma_{ij} \quad , \qquad \overset{(i)}{\sigma}_{ij} = 0$$

On account of (3.18) and (3.19)

$$(4.6) \qquad \sigma_{ij} = \rho\, \frac{\partial f(\varepsilon_{kl}, \theta)}{\partial \varepsilon_{ij}} \quad ,$$

where

H. Ziegler

$$(4.7) \qquad f = u(\mathcal{E}_{kl}, \theta) - \theta s(\mathcal{E}_{kl}, \theta)$$

is the specific free energy. Equation (4.6) is the natural thermodynamic generalization of (4.3). It is obvious that (4.4) also follows from (4.6). Moreover, for isothermal processes, (4.4) and (4.6) reduce to (4.1) and (4.3) respectively. Thus hypoelastic, elastic and hyperelastic materials are identical.

So far we have discussed implications of the fundamental theorems. If the principles of Section 2 are valid, it becomes possible, e.g., to simplify the general constitutive equations established by some authors $\begin{bmatrix} 9, 10, 11 \end{bmatrix}$ for non-newtonian fluids $\begin{bmatrix} 12 \end{bmatrix}$. The rate of work per unit volume is

$$(4.8) \qquad P = \sigma_{jk} V_{jk} = (\overset{(r)}{\sigma}_{jk} + \overset{(i)}{\sigma}_{jk}) V_{jk} \qquad .$$

In a fluid the reversible stress tensor is given by the hydrostatic pressure p alone. Thus

$$(4.9) \qquad \overset{(r)}{\sigma}_{jk} = -p \, \delta_{jk} \qquad .$$

The rate of reversible work is therefore

$$(4.10) \qquad P^{(r)} = \overset{(r)}{\sigma}_{jk} V_{jk} = -p V_{ii} \qquad ,$$

and the rate of dissipation work is given by

$$(4.11) \qquad P^{(i)} = \overset{(i)}{\sigma}_{jk} V_{jk} = \varrho D (V_{jk}) \qquad ,$$

where $\sigma_{jk}^{(i)}$ is the stress tensor due to viscosity and $D(V_{jk})$ denotes the dissipation function per unit mass. Comparing (4.11) with (2.3) we find that, for the unit volume, the $\sigma_{jk}^{(i)}$ are the irreversible forces corresponding to the velocities V_{jk} . Thus the principle of least irreversible force (2.5) requires that

$$(4.12) \qquad \sigma_{jk}^{(i)} = \rho D \left(\frac{\partial D}{\partial V_{lm}} V_{lm} \right)^{-1} \frac{\partial D}{\partial V_{jk}}$$

In an isotropic fluid the dissipation function has the form

$$(4.13) \qquad D\left(V_{(1)}, V_{(2)}, V_{(3)}\right),$$

where

$$(4.14) \qquad \begin{cases} V_{(1)} = V_{ii} \, , \\[2ex] V_{(2)} = \dfrac{1}{2} \left(V_{ij} V_{ji} - V_{ii} V_{jj} \right) , \\[2ex] V_{(3)} = \dfrac{1}{6} \left(2 V_{ij} V_{jk} V_{ki} - 3 V_{ij} V_{ji} V_{kk} + V_{ii} V_{jj} V_{kk} \right) \end{cases}$$

are the basic invariants (see, e.g., $\begin{bmatrix} 7 \end{bmatrix}$, p.22) of the deformation rate. Combining (4.12) and (4.14) , one obtains

$$(4.15) \qquad \sigma_{jk}^{(i)} = K \left[\frac{\partial D}{\partial V_{(1)}} \delta_{jk} + \frac{\partial D}{\partial V_{(2)}} \left(V_{jk} - V_{(1)} \delta_{jk} \right) + \frac{\partial D}{\partial V_{(3)}} \left(V_{ji} V_{ik} - V_{(1)} V_{jk} - V_{(2)} \delta_{jk} \right. \right.$$

H. Ziegler

with

$$(4.16) \qquad K = \rho\, D\Big(\frac{\partial D}{\partial V_{(1)}}\, V_{(1)} + 2\,\frac{\partial D}{\partial V_{(2)}}\, V_{(2)} + 3\,\frac{\partial D}{\partial V_{(3)}}\, V_{(3)}\Big)^{-1} \;.$$

This is the most general constitutive equation of a fluid as defined above. It contains a single, physically significant, function D , while the equations of Reiner, Prager and Rivlin depend on two or three functions without apparent physical meaning.

The constitutive equation (4.15) can be simplified $[12]$ by retaining only certain powers of V_{jk} or by assuming that D depends on a restricted number of fundamental invariants. The simplest special case, obtained by linearization, is Stokes' equation.

If we write the dissipation function in the form (2.6), i. e., in terms of the forces $\sigma_{jk}^{(i)}$, it can be visualized, in stress space (Fig. 2) , by means of the surfaces $D'(\sigma_{jk}^{(i)}) = $ const. On account of (2.7)

$$(4.17) \qquad (\sigma_{jk}^{(i)} - \sigma_{jk}^{(i)*})V_{jk} \gtreqless 0 \quad,$$

where V_{jk} is the actual deformation rate, $\sigma_{jk}^{(i)}$ the actual irreversible stress and $\sigma_{jk}^{(i)*}$ any other irreversible stress with $D'(\sigma_{jk}^{(i)*}) \lesseqgtr D'(\sigma_{jk}^{(i)})$. It follows that the vector V_{jk} in Fig. 2 has the direction of the exterior normal of the D'-surface at P .

The plastic body is obtained as a limiting case by assuming that all D'-surfaces coincide, thus forming the yield surface, which still may depend on the state of the element and on its history. Here (4.17) implies the convexity of the yield surface and supplies v. Mises' theory of the plastic poten-

H. Ziegler

tial $\begin{bmatrix} 13 \end{bmatrix}$. Thus v. Mises' hypothesis, which is fundamental for the theory of plasticity, can be justified by thermodynamic considerations.

There are many more instances where thermodynamics plays an essential part in the formulation of constitutive equations. In the first one of the examples treated here we have merely made use of the fundamental theorems. In the last two cases a more recent and still hypothetical theory has been used. Maybe the results obtained in these examples and in similar cases will contribute to justify this theory.

REFERENCES

1   H. Ziegler, Zwei Extremalprinzipien der irreversiblen Thermodynamik, Ing. Arch. 30, 410 (1961).

2   L. Onsager, Reciprocal Relations in Irreversible Processes, Phys. Rev. 37, II, 405 (1931) and 38, II, 2265 (1931). Compare also H. B. G. Casimir, On Onsager's Principle of Microscopic Reversibility, Rev. mod. Phys. 17, 343 (1945) or S. R. de Groot, Thermodynamics of Irreversible Processes (North-Holland Publishing Co., Amsterdam 1952).

3   C. Truesdell, Reactions of the History of Mechanics upon Modern Research, J. Appl. Mech. 29, Series E, 229 (1962).

4   H. Ziegler, Die statistischen Grundlagen der irreversiblen Thermodynamik, Ing. Arch. 31, 317 (1962).

5   H. Ziegler, Ueber ein Prinzip der grössten spezifischen Entropieproduktion und seine Bedeutung für die Rheologie, Rheol. Acta 2, 230 (1962).

6   H. Ziegler, Some Extremum Principles in Irreversible Thermodynamics, with Application to Continuum Mechanics, in I. N. Sneddon and R. Hill, Progress in Solid Mechanics, vol. IV (North-Holland Publishing Co., Amsterdam), in print.

7   W. Prager, Introduction to Mechanics of Continua (Ginn and Co., Boston 1961).

8   C. Truesdell, The Classical Field Theories, in S. Flügge, Encyclopedia of Physics, vol. III/1 (Springer-Verlag, Berlin 1960), p. 723, 725, 731.

9   M. Reiner, A Mathematical Theory of Dilatancy, Amer. J. Math. 67, 350 (1945).

10 W. Prager, Strain Hardening under Combined Stresses, J. Appl. Phys. 16, 837 (1945).

H. Ziegler

11 <u>R. Rivlin,</u> The Hydrodynamics of Non-Newtonian Fluids I Proc. Roy. Soc. A <u>193</u>, 260 (1948).

12 <u>Ch. Wehrli</u> and <u>H. Ziegler,</u> Einige mit dem Prinzip von der grössten Dissipationsleistung verträgliche Stoffgleichungen, Z. angew. Math. Phys. <u>13</u>, 372 (1962).

13 <u>R. v. Mises,</u> Mechanik der plastischen Formänderung von Kristallen, Z. angew. Math. Mech. 8, 161 (1928).

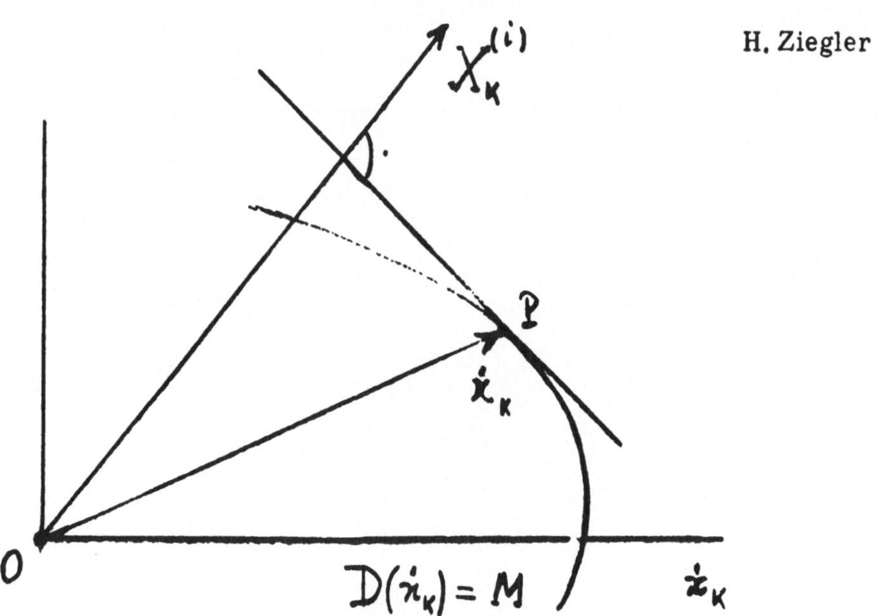

Fig. 1 : Connection between velocity and irreversible force in velocity space.

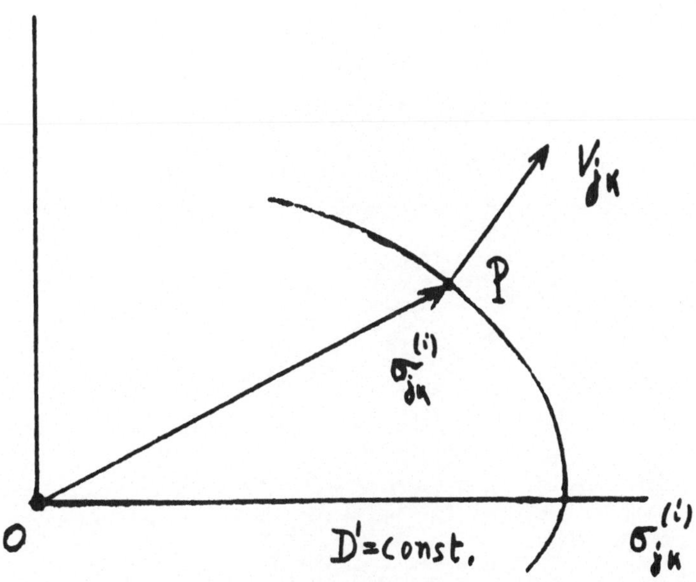

Fig. 2 : Connection between irreversible stress and deformation rate in stress space.

CENTRO INTERNAZIONALE MATEMATICO ESTIVO

(C. I. M. E.)

CATALDO AGOSTINELLI

1. UN TEOREMA DI MEDIA SUL FLUSSO DI ENERGIA
    NEL MOTO DI UN FLUIDO DI ALTA CONDUTTIVITA'
    ELETTRICA IN CUI SI GENERA UN CAMPO MAGNE-
    TICO.

2. SU ALCUNI TEOREMI DI MEDIA IN MAGNETOFLUI-
    DODINAMICA NEL CASO STAZIONARIO.

ROMA - Istituto Matematico dell'Università

UN TEOREMA DI MEDIA SUL FLUSSO DI ENERGIA NEL MOTO DI UN FLUIDO DI ALTA CONDUTTIVITA' ELETTRICA IN CUI SI GENERA UN CAMPO MAGNETICO.

1. Se si considera il moto di un fluido non viscoso, di alta conduttività elettrica, in cui si genera un campo magnetico, sussiste una notevole relazione relativa al flusso di energia totale attraverso una superficie fissa chiusa qualsiasi appartenente al campo del moto del fluido. Se poi gli elementi del campo magnetico e del moto sono periodici rispetto al tempo, si ha l'equivalenza in un periodo del flusso totale di energia attraverso la superficie considerata e del flusso dello <u>stress</u> magnetico e di pressione attraverso la stessa superficie.

2. Le equazioni magnetodinamiche per un fluido perfetto di alta conduttività elettrica, tale da poterla ritenere infinita, scritte nella metrologia gaussiana razionalizzata, si riducono, com'è noto, alle seguenti

$$\frac{\partial \vec{B}}{\partial t} + \text{rot}\,(\vec{B} \wedge \vec{v}) = 0$$

$$\text{div}\,\vec{B} = 0$$

$$\vec{B} = \mu \vec{H}$$

(1)

$$\rho \frac{d\vec{v}}{dt} = \text{rot}\,\vec{H} \wedge \vec{B} - \text{grad}\,p + \rho\,\text{grad}\,U$$

$$\frac{\partial \rho}{\partial t} + \text{div}\,(\rho\vec{v}) = 0$$

$$p = p(\rho),$$

in cui i simboli hanno il solito significato e dove la permeabilità magnetica μ è supposta costante.

Se ora nel campo in cui si muove il fluido consideriamo una superficie chiusa σ qualsiasi, che limita un volume S , moltiplichiamo quindi ambo i

C. Agostinelli

membri dell'equazione del moto scalarmente per il vettore velocità $\vec{v}$ e integriamo sopra tutto il volume S, abbiamo

$$(2) \qquad \int_S \frac{d\vec{v}}{dt} \, x\vec{v} \cdot dS + \int_S \text{grad} p \, x \, \vec{v} \cdot dS - \mu \int_S \text{rot}\vec{H} \wedge \vec{H} \, x \, \vec{v} \cdot dS -$$

$$- \int_S \rho \, \text{grad} \, U \, x \, \vec{v} \cdot dS = 0$$

Tenendo conto dell'equazione di continuità risulta

$$\rho \frac{d\vec{v}}{dt} \, x \, \vec{v} = \frac{1}{2} \rho \frac{dv^2}{dt} = \frac{1}{2} \left[\frac{d}{dt}(\rho v^2) - \frac{d\rho}{dt} v^2 \right] = \frac{1}{2} \left[\frac{\partial}{\partial t}(\rho v^2) + \text{div}(\rho v^2 \cdot \vec{v}) \right]$$

e integrando rispetto al volume S, applicando il teorema della divergenza, si ha

$$(3) \qquad \int_S \rho \frac{d\vec{v}}{dt} \, x \, \vec{v} \cdot dS = \frac{d}{dt} \int_S \frac{1}{2} \rho \, v^2 \cdot dS + \frac{1}{2} \int_\sigma \rho \, v^2 \cdot \vec{v} \, x \, \vec{n} \cdot d\sigma \ ,$$

essendo $\vec{n}$ il versore della normale esterna alla superficie σ .

  Analogamente, essendo la pressione p funzione della densità, se poniamo

$$(4) \qquad \mathcal{P}(p) = \int \frac{dp}{\rho} = \int \frac{1}{\rho} \frac{dp}{d\rho} \, d\rho \ ,$$

possiamo scrivere

$$\text{grad} \, p \, x \, \vec{v} = \rho \, \text{grad} \, \mathcal{P}(\rho) \, x \, \vec{v} = \text{div}\left[\mathcal{P}(\rho) \cdot \rho \, \vec{v} \right] - \mathcal{P}(\rho) \cdot \text{div}(\rho \, \vec{v}) =$$

$$= \text{div}\left[\mathcal{P}(\rho) \cdot \rho \, \vec{v} \right] + \mathcal{P}(\rho) \frac{\partial \rho}{\partial t} \ ;$$

$$\text{ma} \qquad \mathcal{P}(\rho) \frac{\partial \rho}{\partial t} = \frac{\partial}{\partial t}\left[\rho \mathcal{P}(\rho) \right] - \rho \frac{\partial \mathcal{P}}{\partial t} = \frac{\partial}{\partial t}\left[\rho \mathcal{P}(\rho) \right] - \rho \frac{d\mathcal{P}}{d\rho} \frac{\partial \rho}{\partial t} =$$

$$= \frac{\partial}{\partial t}\left[\rho \, \mathcal{P}(\rho) \right] - \frac{\partial p}{\partial t}$$

C. Agostinelli

ne segue

$$\text{grad } p \times \vec{v} = \text{div}\left[\mathcal{P}(\rho).\ \rho\,\vec{v}\right] + \frac{\partial}{\partial t}\left[\rho\,\mathcal{P}(\rho) - p\right]$$

e

$$(5) \qquad \int_S \text{grad } p \times \vec{v}.\,dS = \frac{d}{dt}\int_S\left[\rho\,\mathcal{P}(\rho) - p\right]dS + \oint_\mathfrak{C}\mathcal{P}(\rho).\,\rho\,\vec{v}\times\vec{n}.\,d\sigma,$$

dove $\rho\,\mathcal{P}(\rho) - p$ è l'energia di pressione riferita all'unità di volume.

Nel caso di un fluido in condizioni adiabatiche in cui $p = C\rho^\gamma$, risulta

$$\mathcal{P}(\rho) = \frac{\gamma}{\gamma - 1}\,\frac{p}{\rho}, \qquad \rho\,\mathcal{P}(\rho) - p = \frac{p}{\gamma - 1}.$$

Si ha inoltre, per la prima delle (I)

$$\text{rot}\vec{H} \wedge \vec{H} \times \vec{v} = \text{rot}\vec{H} \times (\vec{H} \wedge \vec{v}) - \vec{H} \times \left[\text{rot}(\vec{H} \wedge \vec{v}) + \frac{\partial \vec{H}}{\partial t}\right] =$$

$$= \text{div}\left[\vec{H} \wedge (\vec{H} \wedge \vec{v})\right] - \frac{1}{2}\frac{\partial H^2}{\partial t},$$

e quindi

$$(6) \qquad \int_S \text{rot}\vec{H} \wedge \vec{H} \times \vec{v}.\,dS = -\frac{d}{dt}\int_S \frac{1}{2}\,H^2.\,dS + \oint_\mathfrak{C}\vec{H} \wedge (\vec{H} \wedge \vec{v}) \times \vec{n}.\,d\sigma.$$

Infine, se il potenziale U delle forze di massa non elettromagnetiche non dipende esplicitamente dal tempo, si ha

$$\rho\,\text{grad } U \times \vec{v} = \text{div}(U.\,\rho\,\vec{v}) - U.\,\text{div}(\rho\,\vec{v}) = \text{div}(U.\,\rho\,\vec{v}) + \frac{\partial}{\partial t}(\rho\,U)$$

e

$$(7) \qquad \int_S \rho\,\text{grad}U \times \vec{v}.\,dS = \frac{d}{dt}\int_S \rho\,U.\,dS + \oint_\mathfrak{C}\rho\,U.\vec{v}\times\vec{n}.\,d\sigma.$$

Sostituendo (3), (5), (6), (7) nella (2) si ottiene

$$\frac{d}{dt}\int_S\left\{\frac{1}{2}\,\rho\,v^2 + \left[\rho\,\mathcal{P}(\rho) - p\right] + \frac{1}{2}\,\mu\,H^2 - \rho\,U\right\}dS +$$

$$+ \oint_\mathfrak{C}\left\{\frac{1}{2}\,\rho\,v^2.\vec{v}\times\vec{n} + \mathcal{P}(\rho).\,\rho\vec{v}\times\vec{n} - \mu\vec{H} \wedge (\vec{H} \wedge \vec{v}) \times \vec{n} - \rho\,U.\vec{v}\times\vec{n}\right\}d\sigma$$

C. Agostinelli

che si può scrivere

$$\frac{d}{dt} \int_S \left\{ \frac{1}{2} \rho v^2 + \left[\rho \mathcal{P}(\rho) - p\right] + \frac{1}{2} \mu H^2 - \rho U \right\} dS +$$

$$(8) \qquad + \oint_\sigma \left\{ \frac{1}{2} \rho v^2 + \left[\rho \mathcal{P}(\rho) - p\right] + \frac{1}{2} \mu H^2 - \rho U \right\} \cdot \vec{v} \times \vec{n} \, d\sigma \; +$$

$$+ \oint_\sigma \left\{ p + \frac{1}{2} \mu H^2 - \mu \mathcal{H}(\vec{H}, \vec{H}) \right\} \vec{v} \times \vec{n} \, d\sigma = 0 \; ,$$

dove $\mathcal{H}(\vec{H}, \vec{H})$ è una diade tale che $\mathcal{H}(\vec{H}, \vec{H}) \vec{v} \times \vec{n} = \vec{H} \times \vec{v} . \vec{H} \times \vec{n}$.

Osserviamo che la quantità

$$(9) \qquad \mathcal{E} = \frac{1}{2} \rho v^2 + \left[\rho \mathcal{P}(\rho) - p\right] + \frac{1}{2} \mu H^2 - \rho U$$

rappresenta l'energia totale riferita all'unità di volume[1], e che l'espressione

ne

$$(10) \qquad \phi = - \left\{ p + \frac{1}{2} \mu H^2 - \mu \mathcal{H}(\vec{H}, \vec{H}) \right\}$$

è lo <u>stress</u> dovuto agli sforzi magnetici e di pressione idrodinamica.

Invero questi sforzi derivano dai termini dell'equazione del moto $\mu \, \text{rot}\vec{H} \wedge \vec{H} - \text{grad}\,p$, che rappresentano l'azione combinata della forza di Lorentz e del gradiente di pressione. Ora risulta

$$\text{rot } \vec{H} \wedge \vec{H} = \frac{d\vec{H}}{dP} \vec{H} - \frac{1}{2} \text{grad } H^2$$

dove $\dfrac{d\vec{H}}{dP}\vec{H}$ è uguale al gradiente della diade $\mathcal{H}(\vec{H}, \vec{H})$ [2]; perciò

---

[1] cfr. C. Agostinelli , <u>Sulla stabilità dei moti magnetofluidodinamici stazionari</u>. "Rendiconti Accad. Naz. dei Lincei", Serie VIII, vol. XXIX, fasc. 6 e vol. XXX, fasc. 1.

[2] Infatti, con riferimento a un sistema di assi cartesiani ortogonali, si ha:

$$\frac{d\vec{H}}{dP} \vec{H} = \sum_i H_i \frac{\partial \vec{H}}{\partial x_i} = \sum_i \frac{\partial}{\partial x_i} (H_i \vec{H}) = \text{grad } \mathcal{H}(\vec{H}, \vec{H})$$

C. Agostinelli

$$\text{rot } \vec{H} \wedge \vec{H} = \text{grad} \left\{ \mathcal{H} \, (\vec{H}, \vec{H}) - \frac{1}{2} H^2 \right\}$$

e quindi

$$(11) \qquad \mu \, \text{rot } \vec{H} \wedge \vec{H} - \text{grad } p = - \text{grad} \left\{ p + \frac{1}{2} \mu H^2 - \mu \, \mathcal{H}(\vec{H}, \vec{H}) \right\} = \text{grad } \phi .$$

L'equazione (8) si può scrivere allora in forma più semplice

$$(12) \qquad \frac{d}{dt} \int_S \mathcal{E} \, dS + \int_{\mathfrak{S}} \mathcal{E} \cdot \vec{v} \times \vec{n} . \, d\sigma = \int_{\mathfrak{S}} \phi \; \vec{v} \times \vec{n} . \, d\sigma$$

che è una notevole relazione relativa al flusso di energia totale attraverso una superficie chiusa qualsiasi.

Ora è noto che il sistema di equazioni (1) ammette delle soluzioni in cui gli elementi del moto e del campo magnetico sono funzioni periodiche rispetto al tempo. Se perciò consideriamo una soluzione periodica di periodo T e integriamo ambo i membri della (12) rispetto a un periodo, poiché risulta $\left[\int_S \mathcal{E} \, dS \right]_0^T = 0$, si deduce

$$(13) \qquad \int_0^T dt \int_{\mathfrak{S}} \mathcal{E} \cdot \vec{v} \times \vec{n} \, d\sigma = \int_0^T dt \int_{\mathfrak{S}} \phi \; \vec{v} \times \vec{n} \, d\sigma$$

la quale esprime il seguente teorema di media :

In un moto periodico di un fluido compressibile, non viscoso, di alta conduttività elettrica, il flusso di energia totale in un periodo attraverso una superficie chiusa qualsiasi fissa, immersa nel campo del moto, è uguale al flusso dello stress magnetico e di pressione attraverso la stessa superficie e nello stesso periodo.

C. Agostinelli

SU ALCUNI TEOREMI DI MEDIA IN MAGNETOFLUIDODINAMICA NEL CASO STAZIONARIO

SUNTO.   Si stabiliscono delle formule che esprimono altrettanti teoremi di di media per i moti magnetofluidodinamici stazionari.

= = = = = =

1. Le equazioni della magnetofluidodinamica, quando si trascura la corrente di spostamento in confronto della corrente di conduzione, col ben noto significato dei simboli sono [1]

$$\operatorname{rot} \vec{H} = \vec{I}$$

$$\operatorname{rot} \vec{E} = -\frac{\partial \vec{B}}{\partial t}$$

$$\operatorname{div} \vec{B} = 0$$

(1)

$$\vec{B} = \mu \vec{H}$$

$$\vec{I} = \gamma (\vec{E} + \vec{v} \wedge \vec{B})$$

$$\rho \frac{d\vec{v}}{dt} = \vec{I} \wedge \vec{B} + \rho \vec{F} - \operatorname{grad} p + (\lambda' + \mu')\operatorname{grad} \operatorname{div} \vec{v} + \mu' \Delta_2 \vec{v}$$

$$\frac{\partial \rho}{\partial t} + \operatorname{div} (\rho \vec{v}) = 0 ,$$

dove la conduttività elettrica γ e la permeabilità magnetica μ , nonchè i coefficienti di viscosità λ', μ', si suppongono costanti.

  Nel caso di moti magnetofluidodinamici stazionari si ha rot $\vec{E}$ = 0,

---

[1] Cfr. C. Agostinelli, Problemi di Magnetofluidodinamica, ecc. "Atti del Simposio sulla Magnetofluidodinamica, Bari 10-14 gennaio 1961" (Edizioni Cremonese, Roma).

C. Agostinelli

e dalle prime cinque delle equazioni (1) si ricava per il campo magnetico l'equazione

$$(2) \qquad \dot{\Delta}_2 \vec{H} = \gamma \mu \ \text{rot} (\vec{H} \wedge \vec{v}) \,.$$

Così pure l'equazione del moto e quella di continuità diventano

$$(3) \qquad \rho \frac{d\vec{v}}{dP} \vec{v} = \mu \frac{d\vec{H}}{dP} \vec{H} - \text{grad} \left(p + \frac{1}{2} \mu H^2\right) + \rho \vec{F} + (\lambda' + \mu') \text{grad div} \vec{v} + $$
$$+ \mu' \Delta_2 \vec{v}$$

$$(4) \qquad \text{div} (\rho \vec{v}) = 0$$

essendo $\dfrac{d\vec{v}}{dP}$ e $\dfrac{d\vec{H}}{dP}$ omografie vettoriali che rappresentano le derivate rispetto al punto P dei vettori $\vec{v}$ ed $\vec{H}$.

2. Ciò premesso consideriamo, nell'interno del campo occupato dal fluido, un dominio sferico S , con centro in un punto qualsiasi P_o , limitato da una superficie sferica σ di raggio r_σ arbitrario, e integriamo ambo i membri della (2) rispetto al volume sferico S . Si ha così

$$(5) \qquad \int_S \Delta_2 \vec{H}. \, dS = \gamma \mu \int_S \text{rot} (\vec{H} \wedge \vec{v}). \, dS \,.$$

Ora, per le formule preliminari di Green, risulta

$$\int_S \Delta_2 \vec{H}. \, dS = \int_\sigma \frac{\partial \vec{H}}{\partial r} \, d\sigma = \frac{d}{dr_\sigma} \int_\sigma \vec{H} d\sigma$$
$$\int_S \text{rot}(\vec{H} \wedge \vec{v}). \, dS = \int_\sigma \vec{n} \wedge (\vec{H} \wedge \vec{v}). \, d\sigma$$

nell'ultima delle quali $\vec{n}$ è il versore della normale esterna alla superficie sferica σ . Sostituendo nella (5) si ottiene

$$(6) \qquad \frac{d}{dr_\sigma} \int_\sigma \vec{H} \, d\sigma = \gamma \mu \int_\sigma \vec{n} \wedge (\vec{H} \wedge \vec{v}). \, d\sigma \,.$$

Applichiamo ora alla sfera S la formula di Green

C. Agostinelli

$$(7) \qquad 4\pi\, U(P_o) = \int_{\mathfrak{S}} \left(\frac{1}{r}\frac{dU}{dn} - U\frac{d\frac{1}{r}}{dn} \right) d\mathfrak{S} - \int_{S} \frac{1}{r}\,\Delta_2 U.\,dS \, ,$$

dove U è funzione finita e continua colle derivate prime e seconde in tutto il dominio che si considera. Essa sussiste anche nel caso in cui U è un vettore, e poichè r è ora la distanza di un punto P dal centro P_o della sfera S, quella formula diventa

$$(7') \qquad 4\pi\, U(P_o) = \frac{1}{r_{\mathfrak{S}}}\frac{d}{dr_{\mathfrak{S}}} \int_{\mathfrak{S}} U\,d\mathfrak{S} + \frac{1}{r_{\mathfrak{S}}^2} \int_{\mathfrak{S}} U\,d\mathfrak{S} - \int_{S} \frac{\Delta_2 U}{r}\,dS \, .$$

Ponendo in luogo di U il vettore $\vec{H}$ che rappresenta il campo magnetico, e tenendo conto della (2), si deduce

$$(8) \qquad 4\pi\, \vec{H}(P_o) = \left(\frac{1}{r_{\mathfrak{S}}}\frac{d}{dr_{\mathfrak{S}}} + \frac{1}{r_{\mathfrak{S}}^2}\right) \int_{\mathfrak{S}} \vec{H}\,d\mathfrak{S} - \gamma\mu\int_{S} \frac{1}{r}\,\mathrm{rot}\,(\vec{H} \wedge \vec{v})\,dS \, .$$

Ma

$$\int_{S} \frac{1}{r}\,\mathrm{rot}(\vec{H}\wedge\vec{v})\,dS = \int_{S} \mathrm{rot}\left(\frac{1}{r}\,\vec{H}\wedge\vec{v}\right)\,dS - \int_{S} \mathrm{grad}\,\frac{1}{r}\wedge(\vec{H}\wedge\vec{v}).\,dS =$$

$$= \frac{1}{r_{\mathfrak{S}}} \int_{\mathfrak{S}} \vec{n}\wedge(\vec{H}\wedge\vec{v})\,d\mathfrak{S} - \int_{S} \mathrm{grad}\,\frac{1}{r}\wedge(\vec{H}\wedge\vec{v}).\,dS \, ,$$

Sostituendo nella (8), e semplificando, avendo riguardo alla (6), si ha infine

$$(9) \qquad 4\pi\,\vec{H}(P_o) = \frac{1}{r_{\mathfrak{S}}^2} \int_{\mathfrak{S}} \vec{H}\,d\mathfrak{S} + \gamma\mu\int_{S} \mathrm{grad}\,\frac{1}{r}\wedge(\vec{H}\wedge\vec{v})\,dS \, ,$$

che esprime un teorema di media per il campo magnetico.

3. Integrando ora ambo i membri dell'equazione (3) del moto rispetto al volume della sfera S, otteniamo

$$(10) \qquad \int_{S} \rho\,\frac{d\vec{v}}{dP}\,\vec{v}.\,dS = \int_{S} \frac{d\vec{H}}{dP}\,\vec{H}.\,dS - \int_{S} \mathrm{grad}\,\left(p + \frac{1}{2}\,\mu\,H^2\right).\,dS + \int_{S} \rho\,\vec{F}.\,dS +$$

$$+ (\lambda' + \mu') \int_{S} \mathrm{grad}\,\mathrm{div}\,\vec{v}.\,dS + \mu' \int_{S} \Delta_2\,\vec{v}.\,dS \, .$$

C. Agostinelli

Avendo riguardo all'equazione (4) di continuità, se si indicano con x_1, x_2, x_3 le coordinate cartesiane del punto P, e con v_1, v_2, v_3 le componenti del vettore $\vec{v}$, risulta

$$\rho \frac{d\vec{v}}{dP} \vec{v} = \sum_{1}^{3} {}_{i} \, \rho \, v_i \, \frac{\partial \vec{v}}{\partial x_i} = \sum_{1}^{3} {}_{i} \, \frac{\partial}{\partial x_i} \, (\rho \, v_i . \vec{v})$$

e quindi, per le formule di Gauss,

$$\int_{S} \rho \, \frac{d\vec{v}}{dP} \, \vec{v} . \, dS = \oint_{\sigma} \rho \, \vec{v} \times \vec{n} . \, \vec{v} . \, d\sigma \quad .$$

Analogamente, essendo div $\vec{H} = 0$, si ha

$$\int_{S} \frac{d\vec{H}}{dP} \, \vec{H} . \, dS = \oint_{\sigma} \vec{H} \times \vec{n} . \, \vec{H} . \, d\sigma \quad .$$

L'equazione (10) diventa pertanto

$$(11) \quad \oint_{\sigma} \rho \vec{v} \times \vec{n} . \, \vec{v} . \, d\sigma = \mu \oint_{\sigma} \vec{H} \times \vec{n} . \, \vec{H} . \, d\sigma - \oint_{\sigma} (p + \frac{1}{2} \mu \, H^2) \vec{n} d\sigma + \int_{S} \rho \vec{F} . \, dS +$$
$$+ (\lambda' + \mu') \oint_{\sigma} \text{div} \, \vec{v} . \, \vec{n} d\sigma + \mu' \frac{d}{dr_{\sigma}} \oint_{\sigma} \vec{v} d\sigma \quad .$$

Applicando d'altra parte la formula (7') di Green al vettore $\vec{v}$, si ottiene

$$(12) \quad 4 \pi \vec{v}(P_0) = (\frac{1}{r_{\sigma}} \frac{d}{dr_{\sigma}} + \frac{1}{r_{\sigma}^2}) \oint_{\sigma} \vec{v} d\sigma - \int_{S} \frac{1}{r} \, \Delta_2 \vec{v} . \, dS \quad .$$

Ma sostituendo in luogo di $\Delta_2 \vec{v}$ il valore che si ricava dalla (3), si ha

$$(13) \quad \int_{S} \frac{1}{r} \, \Delta_2 \vec{v} . \, dS = \int_{S} \frac{1}{r} \left\{ \frac{1}{\mu'} \rho \frac{d\vec{v}}{dP} \vec{v} - \frac{\mu}{\mu'} \frac{d\vec{H}}{dP} \vec{H} + \frac{1}{\mu'} \text{grad} \, (p + \frac{1}{2} \mu \, H^2) - \right.$$
$$\left. - \frac{1}{\mu'} \rho \vec{F} - \frac{\lambda' + \mu'}{\mu'} \, \text{grad div} \, \vec{v} \right\} dS \quad .$$

Con facile calcolo si vede che risulta

$$\int_{S} \frac{1}{r} . \, \rho \, \frac{d\vec{v}}{dP} \vec{v} . \, dS = \frac{1}{r_{\sigma}} \oint_{\sigma} \rho \vec{v} \times \vec{n} . \, \vec{v} . \, d\sigma - \int_{S} \rho \vec{v} \times \text{grad} \frac{1}{r} . \, \vec{v} . \, dS$$

C. Agostinelli

$$\int_S \frac{1}{r} \frac{d\vec{H}}{dP} \, \vec{H}. \, dS = \frac{1}{r_\sigma} \int_\sigma \vec{H} \times \vec{n}. \, \vec{H}. \, d\sigma - \int_S \vec{H} \times \mathrm{grad} \, \frac{1}{r}. \, \vec{H}. \, dS$$

$$\int_S \frac{1}{r} \, \mathrm{grad} \, (p + \frac{1}{2}\mu H^2). \, dS = \frac{1}{r_\sigma} \int_\sigma (p + \frac{1}{2}\mu H^2)\vec{n}. \, d\sigma -$$

$$- \int_S (p + \frac{1}{2}\mu H^2)\mathrm{grad} \, \frac{1}{r} \, . \, dS$$

$$\int_S \frac{1}{r} \, \mathrm{grad} \, \mathrm{div} \, \vec{v}. \, dS = \frac{1}{r_\sigma} \int_\sigma \mathrm{div} \, \vec{v}. \, \vec{n} \, d\sigma - \int_S \mathrm{div} \, \vec{v}. \, \mathrm{grad} \, \frac{1}{r} \, dS \quad .$$

Sostituendo allora nella (13), e quindi nella (12), e semplificando, tenendo conto della (11), si ottiene

$$(14) \quad 4\pi \vec{v}(P_o) = \frac{1}{r_\sigma^2} \int_\sigma \vec{v} d\sigma + \frac{1}{\mu'} \int_S \rho \, \vec{v} \times \mathrm{grad} \, \frac{1}{r}. \, \vec{v}. \, dS -$$

$$- \frac{\mu}{\mu'} \int_S \vec{H} \times \mathrm{grad} \, \frac{1}{r} \, \vec{H}. \, dS + \frac{1}{\mu'} \int_S (p + \frac{1}{2}\mu H^2). \, \mathrm{grad} \, \frac{1}{r}. \, dS +$$

$$+ \frac{1}{\mu'} \left[\int_S \frac{1}{r} \, \rho \, \vec{F} \, dS - \frac{1}{r_\sigma} \int_S \rho \vec{F}. \, dS - \frac{\lambda' + \mu'}{\mu'} \int_S \mathrm{div} \, \vec{v}. \, \mathrm{grad} \, \frac{1}{r}. \, dS \right],$$

che esprime un teorema di media per la velocità in un punto P_o del campo di moto.

Nel caso di un fluido incompressibile di densità ρ_o, e in assenza di forze di natura non elettromagnetica ($\vec{F} = 0$), la (14) diventa più semplicemente

$$(15) \quad 4\pi \vec{v}(P_o) = \frac{1}{r_\sigma^2} \int_\sigma \vec{v} \, d\sigma + \frac{\rho_o}{\mu'} \int_S \vec{v} \times \mathrm{grad} \, \frac{1}{r}. \, \vec{v} \, dS - \frac{\mu}{\mu'} \int_S \vec{H} \times \mathrm{grad} \frac{1}{r}. \, \vec{H} \, dS +$$

$$+ \frac{1}{\mu'} \int_S (p + \frac{1}{2}\mu H^2) \, \mathrm{grad} \, \frac{1}{r} \, . \, dS \quad .$$

C. Agostinelli

4. Osserviamo che la (11), risoluta rispetto all' $\int_{\sigma} (p + \frac{1}{2} \mu H^2) \vec{n}\, d\sigma$, fornisce il risultante della pressione totale sulla superficie σ . Essa è valida
qualunque sia il dominio S , salvo la modifica dell'ultimo integrale, e può
comprendere anche tutto il campo in cui si muove il fluido. In questo caso,
se supponiamo che il fluido elettricamente conduttore sia contenuto in un re-
cipiente limitato da una parete rigida σ , perfettamente conduttrice, al limite su σ dovremo avere

$$\vec{v} \times \vec{n} = 0 \ , \qquad \vec{H} \times \vec{n} = 0 \ ,$$

e in tal caso la (11) porge

$$(16) \quad \int_{\sigma} (p + \frac{1}{2}\mu H^2)\vec{n}\, d\sigma = \int_{S} \rho\, \vec{F}\, dS + (\lambda' + \mu') \int_{\sigma} \mathrm{div}\,\vec{v}.\,\vec{n}\, d\sigma + \mu' \int_{\sigma} \frac{d\vec{v}}{dn}\, d\sigma \ .$$

Se più in particolare il fluido è non viscoso e le forze di massa non elettromagnetiche sono nulle, si ha che il risultante delle pressioni totali sulla superficie limite σ è nullo.

E' facile dimostrare ancora che insieme alla (16) sussiste l'equazione dei
momenti [1]

$$(17) \quad \int_{\sigma} (P-0) \wedge (p + \frac{1}{2}\mu H^2)\vec{n}\, d\sigma = \int_{S} (P-0) \wedge \vec{F}.\, dS +$$

$$+ (\lambda' + \mu') \int_{\sigma} (P-0) \wedge \mathrm{div}\,\vec{v}.\,\vec{n}\, d\sigma + \mu' \left[\int_{\sigma} (P-0) \wedge \frac{d\vec{v}}{dn}\, d\sigma - \int_{\sigma} \vec{n} \wedge \vec{v}.\, d\sigma \right] ,$$

dove 0 è un punto fisso.

---

[1] Si osservi che risulta $(P-0) \wedge \Delta_2\vec{v} = \Delta_2\left[(P-0) \wedge \vec{v}\right] - 2\,\mathrm{rot}\,\vec{v}$, e quindi
$$\int_{S} (P-0) \wedge \Delta_2\vec{v}.\, dS = \int_{\sigma} \frac{d}{dn}\left[(P-0) \wedge \vec{v}\right] d\sigma - 2\int_{\sigma} \vec{n} \wedge \vec{v}.\, d\sigma = \int_{\sigma} (P-0) \wedge \frac{d\vec{v}}{dn}\, d\sigma -$$
$$- \int_{\sigma} \vec{n} \wedge \vec{v}\, d\sigma \ .$$

C. Agostinelli

5. Un altro teorema di media si può stabilire per il vettore vortice. Invero l'equazione (3) del moto si può scrivere

$$\text{rot } \vec{v} \wedge \rho \vec{v} = \mu \text{ rot}\vec{H} \wedge \vec{H} - \text{grad } p - \frac{1}{2} \rho \text{ grad } v^2 + \rho \vec{F} +$$

$$+ (\lambda' + \mu') \text{ grad div } \vec{v} + \mu' \Delta_2 \vec{v}$$

Prendendo il rotore di ambo i membri, e ponendo rot $\vec{v} = \vec{\omega}$, si ricava

$$\Delta_2 \vec{\omega} = \frac{1}{\mu'} \text{rot}(\vec{\omega} \wedge \rho \vec{v}) - \frac{\mu}{\mu'} \text{rot}(\text{rot } \vec{H} \wedge \vec{H}) + \frac{1}{2\mu'} \text{grad} \rho \wedge \text{grad } v^2 - \frac{1}{\mu'} \text{rot}(\rho F).$$

Con procedimento analogo a quello dei numeri precedenti, sempre con riferimento a una sfera S di centro P_o e raggio r_σ , si deduce

$$(18) \quad 4\pi \vec{\omega}(P_o) = \frac{1}{r_\sigma^2} \int_\sigma \vec{\omega} d\sigma + \frac{1}{2\mu' r_\sigma} \int_S \text{grad} \rho \wedge \text{grad } v^2 . dS -$$

$$- \frac{1}{2\mu'} \int_S \frac{1}{r} \text{grad} \rho \wedge \text{grad } v^2 . dS + \frac{1}{\mu'} \int_S \text{grad} \frac{1}{r} \wedge (\vec{\omega} \wedge \rho \vec{v}) dS -$$

$$- \frac{\mu}{\mu'} \int_S \text{grad} \frac{1}{r} \wedge (\text{rot}\vec{H} \wedge \vec{H}) . dS - \frac{1}{\mu'} \int_S \text{grad} \frac{1}{r} \wedge \rho \vec{F} . dS$$

che esprime un teorema di media per il vortice.

Nel caso di un fluido incompressibile e in assenza di forze di massa non elettromagnetiche si ha più semplicemente

$$(19) \quad 4\pi \vec{\omega}(P_o) = \frac{1}{r_\sigma^2} \int_\sigma \vec{\omega} d\sigma + \frac{\rho_o}{\mu'} \int_S \text{grad} \frac{1}{r} \wedge (\vec{\omega} \wedge \vec{v}) dS -$$

$$- \frac{\mu}{\mu'} \int_S \text{grad} \frac{1}{r} \wedge (\text{rot } \vec{H} \wedge \vec{H}) . dS .$$

6. Nel caso di piccoli movimenti di un fluido incompressibile soggetto a un campo magnetico uniforme $\vec{H}_o$, supposto che il campo magnetico indotto $\vec{h}$ sia molto piccolo, trascurando i termini di ordine superiore al primo rispetto a $\vec{v}$ ed $\vec{h}$, e supponendo nulle le forze di massa non elettromagnetiche, le equazioni (2) e (3) si riducono alle seguenti

C. Agostinelli

$$(20) \qquad \Delta_2 \vec{h} = \gamma \mu \, \mathrm{rot}(\vec{H}_0 \wedge \vec{v})$$

$$(21) \qquad \mu \, \mathrm{rot}\vec{h} \wedge \vec{H}_0 - \mathrm{grad}\, p + \mu' \Delta_2 \vec{v} = 0$$

con $\mathrm{div}\, \vec{v} = 0$ e $\mathrm{div}\, \vec{h} = 0$.

Dalle (20) e (21) con procedimento analogo a quello dei numeri precedenti si ricava

$$(22) \qquad 4\pi \vec{h}(P_0) = \frac{1}{r_\sigma^2} \int_\sigma \vec{h}\, d\sigma + \gamma\mu \int_S \mathrm{grad}\frac{1}{r} \wedge (\vec{H}_0 \wedge \vec{v}) dS$$

$$(23) \qquad 4\pi \vec{v}(P_0) = \frac{1}{r_\sigma^2} \int_\sigma \vec{v}\, d\sigma + \frac{1}{\mu'} \int_S p\, \mathrm{grad}\frac{1}{r}\, dS - \frac{\mu}{\mu'} \int_S (\mathrm{grad}\frac{1}{r} \wedge \vec{h}) \wedge \vec{H}_0 \cdot dS$$

Inoltre, prendendo ora la divergenza di ambo i membri della (21) si ha

$$(24) \qquad \Delta_2 (p + \mu \vec{h} \times \vec{H}_0) = 0 \quad ,$$

cioè la $p + \mu \vec{h} \times \vec{H}_0$ è una funzione armonica, e si ha senz'altro

$$(25) \qquad (p + \mu \vec{h} \times \vec{H}_0)_{P_0} = \frac{1}{4\pi r_\sigma^2} \int_\sigma (p + \mu \vec{h} \times \vec{H}_0)\, d\sigma$$

come conseguenza del teorema della media di Gauss.

CENTRO INTERNAZIONALE MATEMATICO ESTIVO

(.C. I. M. E.)

DARIO GRAFFI

1. PRINCIPI DI MINIMO E VARIAZIONALI NEL CAMPO
ELETTROMAGNETICO.
2. TEOREMI DI RECIPROCITA' NEI FENOMENI NON STA-
ZIONARI.

ROMA - Istituto Matematico dell'Università

PRINCIPI DI MINIMO E VARIAZIONALI NEL CAMPO ELETTROMAGNETICO

D. Graffi

1. In questa conferenza non intendo esporre tutti i principi di minimo e variazionali che si incontrano nella teoria del campo elettromagnetico; mi limiterò soltanto a considerare alcuni principi che, a mio avviso, presentano qualche novità.

Comincerò col richiamare due teoremi di minimo per il campo elettrostatico dovuti, in sostanza, a Lord Kelvin.

Si consideri un sistema di n conduttori, elettrizzati, immersi in un dielettrico, che supporremo perfetto, e, solo per semplicità di esposizione, neutro. Il primo teorema di Lord Kelvin afferma che l'energia del campo elettrico compatibile con assegnate cariche sui conduttori è minima in condizioni di equilibrio, cioè quando il campo è elettrostatico. Più precisamente, considerato un campo elettrico $\vec{E}'$ (anche fittizio), infinitesimo all'infinito, col corrispondente vettore spostamento $\vec{D}$ soluzione dell'equazione :

$$(1) \qquad \operatorname{div} \vec{D} = 0$$

e con valori assegnati (uguali alle cariche) per il flusso di $\vec{D}$ attraverso la superficie di ogni conduttore, l'energia che compete al campo $\vec{E}$ diventa minima quando $\vec{E}$ deriva da un potenziale V continuo, costante sulla superficie del conduttore e soddisfacente alle solite condizioni di convergenza all'infinito.

Il secondo teorema di Lord Kelvin afferma che fra tutti i campi elettrici (in generale fittizi) che derivano da un potenziale con valore costante assegnato sui conduttori, e infinitesimo all'infinito, compete l'energia minima al campo elettrostatico, cioè a quello in cui il corrispondente vettore spostamento soddisfa la (1).

181

D. Graffi

I teoremi ora enunciati si provano ammettendo fra vettore spostamento $\vec{D}$ e il campo elettrico $\vec{E}$ (s'intende nello stesso punto dello spazio) la relazione :

$$(2) \qquad \vec{D} = \varepsilon \, \vec{E}$$

dove ε (costante dielettrica) è una grandezza scalare (o al più un tensore doppio simmetrico) funzione, al più, del punto in cui si considerano $\vec{D}$ ed $\vec{E}$. In sostanza la (2) presuppone, come si suol dire, il mezzo lineare. Però, come è noto, in tempi abbastanza recenti sono stati introdotti, anche nella tecnica, dielettrici non-lineari, cioè tali che per essi la (2) viene sostituita da una relazione più complicata non-lineare di cui diremo fra breve. Ci proponiamo di ricavare una estensione dei teoremi di Kelvin ai dielettrici non lineari [1]; dovremo però introdurre le seguenti ipotesi che sembrano plausibili dal punto di vista fisico.

a) L'isteresi sia trascurabile, sicchè il vettore $\vec{D}$ sia funzione univocà,anche non-lineare,del vettore $\vec{E}$. In formule :

$$(3) \qquad \vec{D} = \vec{D}(\vec{E})$$

relazione che generalizza la (2). La funzione $\vec{D}(\vec{E})$ verrà supposta differenziabile, per ogni valore di $\vec{E}$

b) Esista un'energia elettrica, più precisamente, il lavoro del campo elettrico per una variazione $d\vec{D}$ del vettore spostamento sia il differenziale esatto di una funzione scalare (densità di energia del campo elettrico) $p(\vec{E})$ del campo elettrico $\vec{E}$ cioè sia ($\vec{D}$ espresso da (3)) :

$$(4) \qquad \vec{E} \cdot d\vec{D} = d\, p(\vec{E})$$

---

[1] Cfr. D. Graffi - Alcuni teoremi di elettrostatica dei dielettrici non lineari - Scritti matematici in onore di Filippo Sibirani, Bologna, 1957, pag. 143.

D. Graffi

nel caso particolare in cui la (3) si riduca a (2) $\quad d\vec{D} = \varepsilon \, d\vec{E}, \quad p(\vec{E}) = \dfrac{\varepsilon E^2}{2}$

come è ben noto.

c) Sia $\vec{D}(0) = 0$, $p(0) = 0$

d) La $\vec{D}$ sia una funzione crescente di $\vec{E}$. Questa nozione può precisarsi nel modo seguente. Se $\vec{E}_a$ ed $\vec{E}_b$ sono due valori qualsiasi del campo elettrico, $(\vec{E}_a \neq \vec{E}_b)$ e $\vec{D}_a$ e $\vec{D}_b$ i corrispondenti valori del vettore spostamento vale la disuguaglianza :

$$(5) \qquad \left(\vec{E}_b - \vec{E}_a \right) \cdot \left(\vec{D}_b - \vec{D}_a \right) > 0$$

e) La (3) sia invertibile e, per la (5), in un solo modo sicchè :

$$(6) \qquad \vec{E} = \vec{E}(\vec{D})$$

ossia $\vec{E}$ è funzione univoca di $\vec{D}$.

Notiamo che le ipotesi enunciate potrebbero sostituirsi con altre meno restrittive, ma su ciò non insisteremo.

2. Ciò posto passiamo a provare che il primo teorema di Kelvin rimane valido anche in presenza di dielettrici non lineari. Sia $\vec{E}_o$ il campo elettrostatico corrispondente a cariche assegnate sui conduttori e siano Q_r e σ_r rispettivamente la carica e la superficie del conduttore erresimo, il flusso del vettore $\vec{D}_o = \vec{D}(\vec{E}_o)$ attraverso σ_r sarà uguale a Q_r. Sia $\vec{E}_1$ un altro campo elettrico il cui corrispondente valore dello spostamento $\vec{D}_1 = \vec{D}(\vec{E}_1)$ soddisfi la (1) e inoltre il flusso di $\vec{D}_1$ su σ_r valga ancora Q_r.

Confrontiamo ora, in un punto generico del dielettrico, $p(\vec{E}_1)$ con $p(\vec{E}_o)$. A questo scopo osserviamo che per ottenere la differenza $p(\vec{E}_1) - p(\vec{E}_o)$ basterà integrare la (4) facendo variare $\vec{E}$ in modo qualunque. Noi supporremo che $\vec{E}$ vari in modo che $\vec{D}$ soddisfi la relazione :

D. Graffi

dove $\vec{D}' = \vec{D}_1 - \vec{D}_o$ e λ è un parametro variabile fra 0 e 1 sicchè per $\lambda = 0$ $\vec{D} = \vec{D}_o$, per $\lambda = 1$ $\vec{D} = \vec{D}_1$. In altre parole immaginiamo che $\vec{E}$ vari da $\vec{E}_o$ a $\vec{E}_1$ con la legge che si ottiene sostituendo (7) in (6). Si ha così :

$$(8) \qquad d\, p(\vec{E}) = \vec{E}\, d\vec{D} = \vec{E} \cdot \vec{D}'\, d\lambda$$

quindi integrando e poi aggiungendo e togliendo $\vec{E}_o \cdot \vec{D}'$

$$(9) \qquad p(\vec{E}_1) - p(\vec{E}_o) = \int_0^1 E(\vec{D}_o + \lambda \vec{D}') \cdot \vec{D}'\, d\lambda =$$

$$= \vec{E}_c \cdot \vec{D}' + \int_0^1 \overline{E(\vec{D}_o + \lambda \vec{D}') - E(\vec{D}_o)} \cdot \lambda\, \vec{D}'\, d\lambda$$

Ora per la (5) (posto $\vec{D}_a = \vec{D}_o$, $\vec{D}_b = \vec{D}_o + \lambda \vec{D}'$ $\vec{E}_a = \vec{E}(\vec{D}_o)$, $\vec{E}_b = \vec{E}(\vec{D}_o + \lambda \vec{D}')$ si ha che l'ultimo integrale di (9) è positivo quindi :

$$(10) \qquad p(\vec{E}_1) - p(\vec{E}_o) > \vec{E}_o \cdot \vec{D}' \qquad {}^{(2)}$$

Integriamo ora la (10) su tutto il volume v esterno ai conduttori. Si ottiene:

$$(11) \qquad \int_v p(\vec{E}_1)\,dv - \int_v p(\vec{E}_o)\,dv > \int_v \vec{E}_o \cdot \vec{D}'\, dv$$

Ora $\vec{E}_o$ è un campo elettrostatico, quindi $\vec{E}_o = - \operatorname{grad} V$ e V ha sul conduttore erresimo un valore costante che indicheremo con V_r ; inoltre $\vec{E}_o$ è infinitesimo all'infinito del secondo ordine. Quindi tenendo conto che $\vec{D}'$ è pure, per le nostre ipotesi, infinitesimo all'infinito e che $\vec{D}'$, come $\vec{D}_o$ e $\vec{D}_1$, soddisfa la (1) si ha :

$$(11') \qquad \int_v \vec{E}_o \cdot \vec{D}'\, dv = - \int_v \operatorname{grad} V \cdot \vec{D}'\, dv = \int_v \operatorname{div}(V\vec{D}')\,dv = -\sum_1^n V_r \int_{\sigma_r} \vec{D}' \cdot \vec{n}\, d\sigma$$

Ma, come si è visto, i flussi di $\vec{D}_1$ e di $\vec{D}_o$ attraverso σ_r sono gli stessi perciò il flusso di $\vec{D}'$ attraverso la medesima superficie è nullo. Risulta quindi nullo il secondo membro di (11). Allora poichè gli integrali al primo membro di (11) sono rispettivamente l'energia U_1 e U_o del campo $\vec{E}_1$ e del campo elettrostatico $\vec{E}_o$ si ha :

---

(2) Notiamo che se $\vec{E}_o$ è uguale a zero si ha $p(\vec{E}_1) > 0$, cioè la densità di energia è sempre positiva, come del resto è intuitivo.

D. Graffi

(12)
$$U_1 > U_0$$

come dovevasi dimostrare [3].

3. Il secondo teorema di Kelvin si estende anch'esso ai dielettrici non lineari sostituendo all'energia la cosiddetta energia complementare la cui densità $q(E)$ è definita dalla formula :

(13)
$$q(\vec{E}) = \vec{E} \cdot \vec{D} - p(\vec{E})$$

Nel caso lineare, come si verifica facilmente, l'energia complementare coincide con l'energia ordinaria.

Quindi il secondo teorema di Kelvin si esprime mediante la formula:

(14)
$$\int_{V} q(\vec{E}_1)\, dV > \int_{V} q(\vec{E}_0)\, dV$$

Per dimostrare la (14), si osservi che in un punto generico dello spazio si ha :

(15)
$$dq = \vec{E} \cdot d\vec{D} + \vec{D} \cdot d\vec{E} - \vec{E} \cdot d\vec{D} = \vec{D} \cdot d\vec{E}$$

e posto (λ è ancora un parametro variabile tra 0 e 1)

(16)
$$\vec{E} = \vec{E}_0 + \lambda \vec{E}' \qquad\qquad \vec{E}' = \vec{E} - \vec{E}_0$$

si ricava :

(17)
$$q(\vec{E}_1) - q(\vec{E}_0) = \int_{0}^{1} \vec{D}(\vec{E}_0 + \lambda \vec{E}') \cdot \vec{E}'\, d\lambda =$$
$$= \vec{D}_0 \cdot \vec{E}' + \int_{0}^{1} \frac{\vec{D}(\vec{E}_0 + \lambda \vec{E}') - \vec{D}(\vec{E}_0)}{\lambda} \cdot \lambda \vec{E}'\, d\lambda$$

---

[3] Nel testo si è supposto $p(\vec{E}_1)$ nulla, come $p(\vec{E}_0)$, nell'interno del conduttore. Se fosse nel conduttore $p(\vec{E}_1)$ diversa da zero, la U_1 sarebbe uguale all'integrale di $p(\vec{E}_1)$ esteso a tutto lo spazio e la (12) sarebbe valida a maggior ragione.

D. Graffi

Ora ragionando come nel numero precedente si trova che l'integrale è positivo [4].

Quindi integrando la (17) su tutto il volume v esterno ai conduttori si ha :

$$(18) \qquad \int_v q(\vec{E}_1) \, dv - \int_v q(\vec{E}_0) \, dv > \int_v \vec{D}_0 \cdot \vec{E}' \, dv$$

Ma $\vec{E}_0 = -\operatorname{grad} \vec{V}_0$, $E_1 = -\operatorname{grad} \dot{V}_1$, $\vec{E}_1 - \vec{E}_0 = -\operatorname{grad} V'$, $(V' = V_1 - V)$, $\operatorname{div} \vec{D}_0 = 0$. Allora, come nel caso precedente ricordando VD' infinitesimo all'infinito di ordine maggiore di due si ha :

$$(19) \qquad \int_v \vec{E}_c' \cdot \vec{E}' \, dv = - \int_v \operatorname{grad} V' \cdot \vec{D}_0 \, dv = - \int_v dw \, (V' D_c) \, dv = \sum_{1}^{h} r \, V_r' \int_{\sigma_r} \vec{D}_c \cdot \vec{n} \, d\sigma_r$$

e poichè V_1 e V_0 hanno su σ_r lo stesso valore ne segue, sempre su σ_r, V' = 0, quindi il secondo membro di (19) è nullo. Si ha così la (14); l'enunciata estensione del secondo teorema di Kelvin è completamente provata.

Notiamo che i teoremi ora ottenuti potrebbero applicarsi per raggiungere i teoremi di esistenza per le soluzioni delle equazioni dell'elettrostatica non lineare, cioè dell'equazione che si otterrebbe ponendo in (1) la (3) con $\vec{E} = -\operatorname{grad} V$.

Notiamo che teoremi analoghi sono stati ottenuti per il campo magnetostatico in presenza dei corpi ferromagnetici, ma su ciò non insisteremo [5].

---

[4] Supponendo $\vec{E}_0$ e quindi $\vec{D}_0 = 0$ si ha $q(\vec{E}_1) > 0$, cioè l'energia complementare è anch'essa positiva.

[5] D. Graffi - Su una legge del minimo della magnetostatica - Annali Università Ferrara, (7) III (1954) - 25. D. Graffi - Alcuni problemi non-lineari della Fisica matematica - Rendiconti Seminario Matematico dell'Università e del Politecnico di Torino 14 (1954-55) - 75.

D. Graffi

4. Passiamo ora a una questione ben diversa, ma di maggiore interesse pratico; cioè a un principio variazionale a cui soddisfa l'impedenza di una antenna radio.

Si abbia un'antenna formata da un conduttore cilindrico che supporremo perfetto. Al solito immaginiamo che l'azione del sistema eccitatore sull'antenna sia equivalente ad una forza elettromotrice inserita nell'antenna stessa o meglio ad un campo elettrico impresso $\vec{E}_i$ diverso dallo zero in una piccola regione dell'antenna, detta regione di alimentazione.

Ora sia I l'intensità di corrente in una sezione ben determinata nella regione di alimentazione; notiamo che se, come avviene spesso in pratica, quella regione è abbastanza piccola,in essa si può supporre I identica in ogni sezione.

E' noto che l'impedenza Z dell'antenna si può esprimere mediante la formula :

(20)
$$Z = - \int_\sigma \frac{\vec{E} \cdot \vec{j} \, d\sigma}{I^2}$$

dove $\vec{j}$ è la densità di corrente superficiale nell'antenna, $\vec{E}$ il campo elettrico prodotto dalla corrente $\vec{j}$, σ la superficie dell'antenna [6].

Determiniamo ora la proprietà variazionale per la Z a cui si è accennato. Supponiamo di variare la corrente in un'antenna cioè di considerare la corrente $\vec{j} + \delta \vec{j}$, supponiamo però che nella regione di alimentazione sia $\delta \vec{j} = 0$, quindi (I è il flusso attraverso una sezione della re-

---

[6] E' bene notare che sia $\vec{E}$, $\vec{j}$ e I sono vettori o numeri complessi che rappresentano grandezze alternative. E' bene anche notare che essendo l'antenna cilindrica $\vec{j}$ si può supporre parallelo all'asse dell'antenna, e se la sezione dell'antenna è circolare $\vec{j}$ vale in grandezza $\frac{I}{2\pi a}$ (a raggio della sezione).

D. Graffi

gione di alimentazione) rimarrà invariata [7]. Quindi si ha:

$$(21) \qquad \delta Z = - \frac{\delta \int_\sigma \vec{E} \cdot \vec{j} \, d\sigma}{I^2}$$

Ora se $\vec{j}$ varia di $\delta \vec{j}$, $\vec{E}$ varia di $\delta \vec{E}$. Sarà perciò a meno di infinitesimi:

$$(22) \qquad \delta \int \vec{E} \cdot \vec{j} \, d\sigma = \int_\sigma \vec{E} \cdot \delta \vec{j} \, d\sigma + \int_\sigma \delta \vec{E} \cdot \vec{j} \, d\sigma$$

Ora si noti che $\delta \vec{E}$ vale il campo generato dalla corrente $\delta \vec{j}$; per il teorema di reciprocità del campo elettromagnetico si può scrivere :

$$(23) \qquad \int_\sigma \delta \vec{E} \cdot \vec{j} \, d\sigma = \int_\sigma \vec{E} \cdot \delta \vec{j} \, d\sigma$$

Ora σ si può dividere in due parti, la regione di alimentazione σ ' e la regione rimanente σ ''. Nella prima $\delta \vec{j}$ = 0; nella seconda, per le proprietà dei conduttori perfetti, e poichè il campo impresso è nullo, $\vec{E}$ è normale alla superficie dell'antenna cioè $\vec{E} \cdot \delta \vec{j}$ = 0; quindi i termini al secondo membro di (22) sono nulli e si ha la relazione cercata :

$$(24) \qquad \delta Z = 0$$

Questo risultato è importante anche dal punto di vista pratico. Poichè la distribuzione della corrente in antenna è nota solo in modo approssimato potremo calcolare Z solo assumendo un valore approssimato della corrente che differirà dal valore esatto per termini dell'ordine di η. Per la (24) considerando η come infinitesimo si deduce che nel calcolo di Z si commette un errore dell'ordine di η^2, sicchè la formula (20) è la più adatta per il calcolo dell'impedenza dell'antenna.

---

[7] Non sarà inutile notare che se I è identica nella regione di alimentazione e se l'antenna è a sezione circolare per quanto si è detto nella nota [6] l'ipotesi $\delta \vec{j}$ = 0 equivale a supporre invariata la I .

D. Graffi

TEOREMI DI RECIPROCITA' NEI FENOMENI NON STAZIONARI

1. E' ben noto il teorema di reciprocità di Betti nella elastostatica ordinaria. Esso si esprime mediante la formula :

$$(1) \qquad \int_v \rho \vec{F}' \cdot \vec{s}'' \, dv + \int_\sigma \vec{f}' \cdot \vec{s}'' \, d\sigma = \int_v \rho \vec{F}'' \cdot \vec{s}' \, dv + \int_\sigma \vec{f}'' \cdot \vec{s}' \, dv$$

dove v indica il volume del corpo elastico, σ la superficie che lo limita, $\vec{s}'$ ed $\vec{s}''$ sono gli spostamenti dovuti, rispettivamente, alle forze di massa e superficiali, $\rho \vec{F}'$, $\vec{f}'$, $\rho \vec{F}''$, $\vec{f}''$ (ρ densità) agenti sul corpo.

Il teorema espresso dalla (1), di notevole importanza pratica, è in certo modo, il capostipite di una serie di teoremi di reciprocità validi in diversi campi della fisica matematica.

Ora è ovvio che il teorema di Betti si può estendere alla elastodinamica aggiungendo nella (1) alle $\rho \vec{F}'$, $\rho \vec{F}''$ rispettivamente i termini

$$-\rho \frac{\partial^2 \vec{s}'}{\partial t^2} \;, \; -\rho \frac{\partial^2 \vec{s}''}{\partial t^2}$$

(t indica il tempo) cioè le forze d'inerzia cambiate di segno. Si ottiene però in tal modo una relazione, di poco interesse pratico, perchè in essa compaiono termini, le forze d'inerzia, che sono, in generale, incogniti. E' perciò naturale domandarsi se nell'elastodinamica vale, almeno sotto certe condizioni, un teorema analogo a quello espresso da (1), ma in cui non compaiono termini d'inerzia. Alcuni risultati in quest'ordine di idee sono stati da me ottenuti applicando la trasformazione di Laplace (1). In questa conferenza intendo ritrovare i detti risultati evitando la trasformazione di Laplace (essa introduce infatti ipotesi sul comportamento per t tendente all'infini-

---

(1) D. Graffi "Sui teoremi di reciprocità nei fenomeni dipendenti dal tempo" Annali di Matematica (IV) XVIII, (1939), 173. "Sul teorema di reciprocità nella dinamica dei corpi elastici" Memorie Accademia Scienze Bologna (10)

./.

D. Graffi

to degli spostamenti che almeno, dal punto di vista concettuale, è opportuno evitare ma, mediante un procedimento adoperato da Gurtin e Sternberg [2] nei loro studi sulla viscoelasticità. Inoltre farò vedere, con un esempio preso dalla meccanica ondulatoria, come quel procedimento possa condurre e relazioni di reciprocità valide in altri campi della fisica e per fenomeni non stazionari.

2. Converrà, per ottenere la preannunciata estensione della (1)[3] scrivere le equazioni della elastodinamica quando sul corpo agiscono le forze di massa e superficiali $\rho \vec{F}'(\tau)$, $\vec{f}'(\tau)$ funzioni, oltre che del luogo, anche del tempo che, come vedremo in seguito, converrà indicare ora con τ . Si ha così, con le solite notazioni tensoriali [4], indicando con $p'_{ij}(\tau)$ il corrispondente tensore degli sforzi all'istante τ e in un punto generico del corpo, con $\rho F'_i(\tau)$, $f'_i(\tau)$ le componenti (covarianti) delle forze di massa e superficiali, con n^i le componenti (contravarianti) del versore normale $\vec{n}$ a σ e diretto verso l'esterno del volume v :

$$(2) \quad \rho F'_i(\tau) + p'_{ij}(\tau)^{\,lj} = \rho \, \frac{\partial^2 s'_i(\tau)}{\partial \tau^2} \qquad\qquad (3) \quad p'_{ij}(\tau)\, n^j - f'_i(\tau)$$

./. IV (1946-47), 103. Über den Reziprositätsatz in der Dynamik der Elastischen Körpen - Ingenieur Archiv 22, 1954, 45.

[2] M. E. Gurtin e Eli Sternberg - On the linear theory of viscoelasticity - Archive of Rational Mechanics and Analysis, 11 (1962) 291.

[3] Ammetteremo che le forze, gli spostamenti e le derivate prime e seconde di queste ultime grandezze soddisfino le solite condizioni di regolarità della Fisica matematica.

[4] Cfr. per esempio B. Finzi ed M. Pastori - Calcolo tensoriale - Zanichelli Bologna, cap. IX.

D. Graffi

Ovviamente la (2) vale in ogni punto di v , la (3) in ogni punto di σ

Ora fra il tensore di deformazione :

$$(4) \qquad \xi'_{ij}(\tau) = \frac{s'_{i/j} + s'_{j/i}}{2}$$

e il tensore degli sforzi passa la relazione di Hooke :

$$(5) \qquad p'_{ij}(\tau) = C_{ijrs}\, \xi'^{rs}(\tau)$$

dove i coefficienti C_{ijrs} sono costanti o al più funzioni solo del posto. I-
noltre ammessa l'esistenza del potenziale elastico si ha $C_{ijrs} = C_{rsij}$.

Ciò posto, se t è un istante superiore a τ , moltiplichiamo la (2) per
$s''^{i}(t-\tau)$ $(s''^{i}(t))$ componente controvariante dello spostamento $\vec{s}''(\tau)$
generato dalle forze $\rho\,\vec{F}''(\tau)$, $\vec{f}''(\tau)$; ovviamente $\vec{s}''(\tau)$ soddisfa le
equazioni analoghe alle (2), (3), (5)). Sommiamo rispetto a i e integriamo
sul volume v . Si ha :

$$(6) \qquad \int_v \rho\, F'_i(\tau)\, s''^{i}(t-\tau)\,dv + \int_v p'_{ij}{}^{/j}(\tau)\, s''^{i}(t-\tau)\,dv = \int_v \rho\, \frac{\partial^2 s'_i(\tau)}{\partial \tau^2}\, s''^{i}(t-\tau)\,dv$$

Ma :

$$(7) \qquad p'_{ij}(\tau)s''^{i}(t-\tau) = \left(p'_{ij}(\tau)s''^{i}(t-\tau) \right)^{/j} - p'_{ij}(\tau)s''^{i/j}(t-\tau)$$

Sostituendo nella (6), applicando il teorema della divergenza e ricordando la
(3) si ha :

$$(8) \qquad \int_v \rho\, F'_i(\tau)\, s''^{i}(t-\tau)\,dv + \int_\sigma f'_i(\tau)\, s''^{i}(t-\tau)\,d\sigma = \int_v p'_{ij}(\tau)s''^{i/j}(t-\tau)\,dv +$$

$$+ \int_v \rho\, \frac{\partial^2 s'_i(\tau)}{\partial \tau^2}\, s''^{i}(t-\tau)\,dv$$

Integrando ora la (8) da 0 a t e osservando che [5] :

---

[5] Il simbolo $\cdot$ fra due vettori è il notissimo simbolo di prodotto scalare.

191

D. Graffi

$$(8') \qquad \rho\, F_i'(\tau)\, s''^i(t-\tau) = \rho\, \vec{F}'(\tau)\cdot\vec{s}''(t-\tau) \qquad f_i(\tau)\, s''^i(t-\tau) = \vec{f}'(\tau)\cdot\vec{s}''(t-\tau)$$

si ottiene con ovvia inversione dell'ordine di integrazione :

$$(9) \quad \int_{?}^{t} d\tau \int_{v} \rho\, \vec{F}'(\tau)\cdot\vec{s}''(t-\tau)\, dv + \int_{0}^{t} d\tau \int_{\sigma} \vec{f}'(\tau)\cdot\vec{s}''(t-\tau)\, d\sigma =$$

$$= \int_{v} dv \int_{0}^{t} P_{ij}'(\tau)\, s''^{ij}(t-\tau)\, d\tau + \int_{v} dv \int_{0}^{t} \rho\, \frac{\partial^2 s_i'(\tau)}{\partial\tau^2}\, s''^i(t-\tau)\, d\tau$$

Ora si ha integrando per parti :

$$(10) \quad \int_{0}^{t} \frac{\partial^2 s_i'(\tau)}{\partial\tau^2}\, s''^i(t-\tau)\, d\tau = \frac{\partial s_i'(t)}{\partial t}\, s''^i(0) - \frac{\partial s_i'(0)}{\partial t}\, s''^i(t) - \int_{0}^{t} \frac{\partial s_i'(\tau)}{\partial\tau}\, \frac{\partial s''^i(t-\tau)}{\partial\tau}\, d\tau$$

Si ha poi ricordando che $P_{ij}'(\tau)$ è un tensore simmetrico :

$$(11) \quad P_{ij}'(\tau)\, s''^{ij}(t-\tau) = P_{ij}'(\tau)\, \mathcal{F}''^{ij}(t-\tau) = C_{ijrs}\, \mathcal{F}'^{rs}(\tau)\, \mathcal{F}''^{ij}(t-\tau)$$

sostituendo nella (9) si ottiene (facendo anche uso di formule analoghe a (8):

$$(12) \quad \int_{0}^{t} d\tau \left(\int_{v} \rho\vec{F}'(\tau)\cdot\vec{s}''(t-\tau)\, dv + \int_{\sigma} \vec{f}'(\tau)\cdot\vec{s}''(t-\tau)\, d\sigma \right) - \int_{v} \rho \left(\frac{\partial\vec{s}'(t)}{\partial t}\cdot\vec{s}''(0) - \right.$$

$$\left. - \frac{\partial\vec{s}'(0)}{\partial t}\cdot\vec{s}''(t) \right) dv = \int_{v} dv \int_{0}^{t} C_{ijrs}\, \mathcal{F}'^{rs}(\tau)\, \mathcal{F}''^{ij}(t-\tau)\, d\tau - \int_{v} \rho\, dv \int_{0}^{t} \frac{\partial s_i'(\tau)}{\partial\tau}\, \frac{\partial s''^i(t-\tau)}{\partial\tau}\, d\tau$$

Ora ricordiamo che se $A(t)$ e $B(t)$ sono funzioni integrabili vale la relazione [6] :

$$(13) \quad \int_{0}^{t} A(\tau)\, B(t-\tau)\, d\tau = \int_{0}^{t} A(t-\tau)\, B(\tau)\, d\tau$$

relazione che si ricava, assai facilmente, eseguendo al primo membro di (13) il cambiamento di variabili $v = t - \tau$ e poi cambiando v con τ . In modo analogo si ha :

$$(13') \quad \int_{0}^{t} \frac{dA(\tau)}{d\tau}\, B(t-\tau)\, d\tau = -\int_{0}^{t} \frac{dA(t-\tau)}{d\tau}\, B(\tau)\, d\tau$$

$$(13'') \quad \int_{0}^{t} \frac{dA(\tau)}{d\tau}\, \frac{dB(t-\tau)}{d\tau}\, d\tau = \int_{0}^{t} \frac{dA(t-\tau)}{d\tau}\, \frac{dB(\tau)}{d\tau}\, d\tau$$

---

[6] Cfr. per esempio: Ghizzetti, Calcolo simbolico, Zanichelli, Bologna 1944 pag. 30.

D. Graffi

Tornando alla (12) si ha (in base alla (13) e (13')) scambiando la posizione degli apostrofi che il secondo membro invariato (nel termine tensoriale occor-
re scambiare gli indici r, s con i, j e tener presente che $C_{ijrs} = C_{rsij}$).
Perciò scambiando gli apostrofi rimane invariato anche il primo membro di
(12) e si ha così la relazione :

$$(14) \quad \int_0^t d\tau \left(\int_V \rho \vec{F}'(\tau) \cdot \vec{a}''(t-\tau)\, dv + \int_S \vec{f}'(\tau) \cdot \vec{a}''(t-\tau)\, d\sigma \right) - \int_V f \left(\frac{\partial \vec{a}'(t)}{\partial t} \cdot \vec{a}''(0) - \frac{\partial \vec{a}''(0)}{\partial t} \cdot \vec{a}'(t) \right) dv$$

$$= \int_0^t d\tau \left(\int_V \rho \vec{F}''(\tau) \cdot \vec{a}'(t-\tau)\, dv + \int_S \vec{f}''(\tau) \cdot \vec{a}'(t-\tau)\, d\sigma \right) \cdot \int_V f \left(\frac{\partial \vec{a}''(t)}{\partial t} \cdot \vec{a}'(0) - \frac{\partial \vec{a}'(0)}{\partial t} \cdot \vec{a}''(t) \right) dv$$

E questa relazione esprime l'estensione alla dinamica del teorema di Betti.

3. Indichiamo alcuni casi particolari della (14): Supponiamo le condizioni ini-
ziali nulle cioè $\vec{a}'(0) = \vec{a}''(0) = 0$, $\dfrac{\partial \vec{a}'(0)}{\partial t} = \dfrac{\partial \vec{a}''(0)}{\partial t} = 0$.
Supponiamo poi

$$(15) \qquad \rho \vec{F}'(t) = G(t)\vec{a}' \qquad\qquad \vec{f}'(t) = G(t)\vec{b}'$$
$$\rho \vec{F}''(t) = G(t)\vec{a}'' \qquad\qquad \vec{f}''(t) = G(t)\vec{b}''$$

dove a', a'', b', b'' sono vettori funzioni del posto, ma non del tempo. In altre parole supponiamo che le forze si possano scomporre nel prodotto di una
funzione del posto per una funzione del tempo; la funzione del tempo identica
per tutte le forze. Si ha allora da (14)

$$(16) \quad \int G(t)\, d\tau \left(\int_V \rho \vec{a}' \, a''(t-\tau)\, dv + \int_S \vec{b}' \, a''(t-\tau)\, d\sigma \right) = \int_0^t G(\tau)\, d\tau \left(\int_V \rho \vec{a}'' \cdot a'(t-\tau)\, dv + \int_S \vec{b}'' \cdot a'(t-\tau)\, d\sigma \right)$$

Ora è noto che se per ogni t :

$$(17) \qquad \int_0^t A(\tau)\, B(t-\tau)\, d\tau = 0$$

e se A(t) non è identicamente nulla, ne segue $B(t) \equiv 0$ [7] quindi dalla (14) si

---

[7] Un modo relativamente semplice per provare questo teorema è il seguente.
Sia T un numero positivo; si ponga $A^*(t) = A(t)$ per t < T, $A^*(t) = 0$ per

D. Graffi

ha :

$$(18) \quad \int_V \rho \vec{a}' \cdot \vec{s}''(t)\, dv + \int_\sigma \vec{b}' \cdot \vec{s}''(t)\, d\sigma = \int_V \rho \vec{a}'' \cdot \vec{s}'(t)\, dv + \int_\sigma \vec{b}'' \cdot \vec{s}'(t)\, d\sigma$$

e moltiplicando per G(t) si trova che la (1), cioè il teorema di Betti nella forma originaria, vale, istante per istante, purchè le condizioni iniziali siano nulle e le forze soddisfino le (15).

Supponiamo ancora, più in particolare, che il primo sistema di forze si riduca rispettivamente ad una forza $\vec{R}(A, t)$ concentrata in A , il secondo a una forza $\vec{R}(B, t)$ concentrata in B e che le due forze mutino col tempo in modo identico. Supposto, per fissare le idee, A e B interni al corpo, le forze superficiali $\vec{f}'$ ed $\vec{f}''$ saranno nulle e si potrà scrivere :

$$\rho \vec{F}'(t) = \delta(P-A)\, \vec{R}(A, t) \qquad\qquad \vec{R}'(A, t) = G(t)\, \vec{m}'(A)$$

$$\rho \vec{F}''(t) = \delta(P-B)\, \vec{R}''(A, t) \qquad\qquad \vec{R}''(A, t) = G(t)\, \vec{m}''(B)$$

dove $\vec{m}'$ ed $\vec{m}''$ sono vettori costanti, $\delta(P-A)$ è la funzione di Dirac del punto generico P . Sostituendo in (14) e indicando con $\vec{s}'(A, t)$, $\vec{s}''(B, t)$ gli spostamenti $\vec{s}''$ ed $\vec{s}'$ all'istante t e rispettivamente in A e B , si ottiene (scambiando t, con t-τ, tenendo presente (13) e ricordando inoltre le proprietà della funzione di Dirac) :

$$\int_0^t G(t-\tau)\left(\vec{m}'(A) \cdot \vec{s}''(A, \tau) - \vec{m}''(B) \cdot \vec{s}'(B, t)\right) d\tau = 0$$

./. t > T e analoga definizione sia valida per $B^*(t)$, T si può scegliere in modo arbitrario purchè A(t) non sia identicamente nulla per t < T. Poichè la trasformata di Laplace $\mathcal{L}(A^*(t))$, $\mathcal{L}(B^*(t))$ sono assolutamente convergenti si ha (cfr. Ghizzetti loc. cit. pag. 33)

$$\mathcal{L}\int_0^t A^*(\tau)\, B^*(t-\tau)\, d\tau = \mathcal{L}(A^*(t))\, \mathcal{L}(B^*(t))$$

da cui poichè A (t) non è identicamente nulla ne segue $\mathcal{L}(B^*(t)) = 0$ quindi $B^*(t) \equiv 0$ perciò per t < T $B(t) \equiv 0$. Dall'arbitrarietà di T segue $B(t) \equiv 0$ per ogni t.

D. Graffi

e per quanto si è affermato poco fa :

$$(19) \qquad \vec{R}'(A,t) \cdot \vec{\Delta}(A,t) = \vec{R}''(B,t) \cdot \vec{\Delta}'(B,t)$$

relazione che estende alla elastodinamica il teorema di reciprocità di Maxwell.

Indichiamo un'altra conseguenza della (14). Siano nulle le forze superficiali, le forze di massa soddisfino le (15) inoltre sia :

$$(20) \qquad \begin{aligned} \vec{\Delta}'(o) &= \lambda \vec{a}' & \frac{\partial \vec{\Delta}'(o)}{\partial t} &= \mu \vec{a}' \\[2mm] \vec{\Delta}''(o) &= \lambda \vec{a}'' & \frac{\partial \vec{\Delta}''(o)}{\partial t} &= \mu \vec{a}'' \end{aligned}$$

dove λ e μ sono costanti. La (14) si può scrivere in questo caso tenendo presente (13)

$$(21) \quad \int_o^t G(t-\tau)\,d\tau \int_v \left(\rho \vec{a}' \cdot \vec{\Delta}''(\tau) - \rho \vec{a}'' \cdot \vec{\Delta}'(\tau) \right) d\tau - \frac{d}{dt} \int_v \rho \left(\vec{\Delta}'(t) \cdot \vec{\Delta}''(o) - \vec{\Delta}'(o) \vec{\Delta}''(t) \right) dv +$$

$$+ \int_v \rho \left(\frac{\partial \vec{\Delta}'(o)}{\partial t} \cdot \vec{\Delta}''(t) - \frac{\partial \vec{\Delta}''(o)}{\partial t} \cdot \vec{\Delta}'(t) \right) dv = o$$

Ossia ricordando (20)

$$(22) \quad \int_o^t G(t-\tau)\,d\tau \int_v \left(\rho \vec{a}' \cdot \vec{\Delta}''(t) - \rho \vec{a}'' \cdot \vec{\Delta}'(t) \right) dv + \lambda \frac{d}{dt} \int_v \rho \left(\vec{a}' \cdot \vec{\Delta}''(t) - \vec{a}'' \cdot \vec{\Delta}'(t) \right) dv +$$

$$+ \mu \int_v \rho \left(\vec{a}' \cdot \vec{\Delta}''(t) - \vec{a}'' \cdot \vec{\Delta}'(t) \right) dv = o$$

Quindi per le proprietà delle equazioni integro differenziali di Volterra si ottiene

$$(23) \qquad \int_v \rho \vec{F}''(t) \cdot \vec{\Delta}'(t)\, dv = \int_v \rho \vec{F}'(t) \cdot \vec{\Delta}''(t)\, dv$$

Cioè la (1) può essere valida in elastodinamica anche quando le condizioni iniziali non sono nulle.

Per altre conseguenze della (14) rimando alle memorie citate.

D. Graffi

4. Come si è detto in principio applichiamo ora le considerazioni precedenti all'equazione temporale di Schrödinger [8]:

$$(24) \qquad \Delta \psi - \frac{8\pi^2 m}{h^2} U \psi = \frac{-4\pi i m}{h} \frac{\partial \psi}{\partial \tau}$$

in cui i simboli hanno il ben noto significato, in particolare U è il potenziale indipendente da τ. Siano $\psi'(\tau)$ e $\psi''(\tau)$ due soluzioni della (24) corrispondenti a diversi valori iniziali. Si consideri l'espressione :

$$(25) \qquad \psi''(t-\tau)\, \mathrm{div\,grad}\, \psi'(\tau) - \frac{8\pi^2 m}{h^2} U \psi'(\tau)\psi''(t-\tau) =$$
$$= - \frac{4\pi i m}{h} \frac{\partial \psi'(\tau)}{\partial \tau} \psi''(t-\tau)$$

Integriamo il primo termine al primo membro di (25) su tutto lo spazio. Si ottiene (v indica ora l'estensione dello spazio infinito) :

$$(26) \qquad \int_v \psi''(t-\tau)\, \mathrm{div\,grad}\, \psi'(\tau)\, dv = \int_v \mathrm{div}\left(\psi''(t-\tau)\, \mathrm{grad}\, \psi'(\tau)\right) dv -$$
$$- \int_v \left(\mathrm{grad}\, \psi''(t-\tau) \cdot \mathrm{grad}\, \psi'(\tau)\right) dv$$

Ora se, come si ammette di solito, $\psi''(t-\tau)\, \mathrm{grad}\, \psi'(\tau)$ è infinitesimo all'infinito di ordine maggiore di due, il primo termine al secondo membro di (26) è nullo. Si integri ora la (25) da 0 a t e poi su tutto il volume dello spazio. Tenendo presente la (26) si ha :

$$(27) \qquad -\int_v dv \int_0^t \mathrm{grad}\, \psi'(t-\tau) \cdot \mathrm{grad}\, \psi'(\tau)\, d\tau - \frac{8\pi^2 m}{h^2}\int_v U\, dv \int_0^t \psi'(\tau)\psi''(t-\tau)\, d\tau =$$
$$= \frac{-4\pi i m}{h}\int_v dv \int_0^t \frac{\partial \psi'(\tau)}{\partial \tau}\, \psi''(t-\tau)\, d\tau$$

Ora cambiando la posizione degli apostrofi si ha, ricordando (13), che il primo membro di (27) non cambia. Si ottiene così la relazione :

---

[8] Cfr. E. Persico, Fondamenti della meccanica atomica, Zanichelli, Bologna, 1936, pag. 169.

D. Graffi

$$(28) \quad \int_V dv \int_0^t \frac{\partial \psi'(\tau)}{\partial \tau} \psi''(t-\tau)\, d\tau = \int_V dv \int_0^t \frac{\partial \psi''(\tau)}{\partial \tau} \psi'(t-\tau)\, d\tau$$

Ora integrando per parti si ha :

$$(29) \quad \int_0^t \frac{\partial \psi'(\tau)}{\partial \tau} \psi''(t-\tau)\, d\tau = \psi'(t)\,\psi''(0) - \psi'(0)\,\psi''(t) - \int_0^t \psi'(\tau)\,\frac{\partial \psi''(t-\tau)}{\partial \tau}\, d\tau$$

sostituendo in (28) e tenendo presente (13') si ha dopo semplici calcoli :

$$(30) \quad \int_V \psi'(t)\,\psi''(0)\, dv = \int_V \psi'(0)\,\psi''(t)\, dv$$

che è la cercata relazione di reciprocità.

Come applicazione della (30) indichiamo il seguente teorema di cui, per brevità, ometteremo la dimostrazione. Se da una certa misura risulta che un corpuscolo è inizialmente nel punto A , la probabilità di trovarlo all'istante t nel punto B , vale la probabilità di trovarlo, nello stesso istante, in A qualora risultasse inizialmente in B .

CENTRO INTERNAZIONALE MATEMATICO ESTIVO

(C. I. M. E.)

G. GRIOLI

I : PROPRIETA' GENERALI DI MEDIA NELLA MEC-
CANICA DEI CONTINUI E LORO APPLICAZIONI.

II : PROBLEMI DI INTEGRAZIONE NELLA TEORIA
DELL'EQUILIBRIO ELASTICO.

ROMA - Istituto Matematico dell'Università

PROPRIETA' GENERALI DI MEDIA
NELLA MECCANICA DEI CONTINUI E LORO APPLICAZIONI

GIUSEPPE GRIOLI

Quanto dirò ha lo scopo di stabilire alcune limitazioni per lo stato tensionale di un corpo continuo e - quando possibile - per la sua deformazione. Le limitazioni stabilite per lo stress mantengono la loro validità qualunque sia la natura del corpo continuo, anche rigido, ma la loro concreta applicabilità presuppone la conoscenza dello stato attuale. Se questo è noto, sia nel caso statico come pure in quello dinamico, è noto il campo d'integrazione che in modo essenziale interviene nelle espressioni delle limitazioni stabilite ed esse acquistano significato concreto quando siano conosciute tutte le forze esterne, in esse comprese quelle d'inerzia.

Pertanto, dalle considerazioni svolte nei primi tre paragrafi non esula neppure il caso delle deformazioni finite se la configurazione attuale - o di equilibrio - è nota.

Vi rientra sempre il caso dei corpi poco deformabili, in quanto per essi nella valutazione degli integrali il campo d'integrazione va classicamente sostituito da quello di una vicinissima configurazione nota di riferimento nei cui punti si ritengono applicate, senza loro alterazione, le forze esterne.

Le difficoltà che nascono se una parte della sollecitazione esterna superficiale non è nota in quanto abbia carattere di reazione vincolare sono superabili in base al contenuto dell'Osservazione riportata alla fine del quarto paragrafo.

G. Grioli

1. Premesse di carattere generale

Denoterò con $\mathcal{C}$ la configurazione attuale di un qualunque corpo continuo riferito ad una terna trirettangola levogira $O\,x_1\,x_2\,x_3$, con

$X_{rs} = X_{sr}$ le sei caratteristiche dello stress, funzioni del generico punto P di $\mathcal{C}$, con μ la densità in P e con μF_r il vettore che esprime la forza di massa specifica, compresa in essa la forza d'inerzia (caso dinamico). Detto Σ il contorno completo di $\mathcal{C}$, sia $f_r d\Sigma$ la forza superficiale esterna agente su $\mathcal{C}$ attraverso l'elemento superficiale $d\Sigma$ e N_r il vettore unitario normale a Σ e orientato verso l'interno di $\mathcal{C}$.

Le equazioni fondamentali sono

$$X_{rs}^{\ /s} = \mu F_r \ , \qquad\qquad (\text{in }\mathcal{C}) \ ,$$

(1)

$$X_{rs} N^s = f_r \ , \qquad\qquad (\text{su }\Sigma) \ ,$$

ove la sbarretta denota derivazione rispetto alla x_s e vale la consueta convenzione della somma.

Per ogni scelta dei numeri interi, positivi o nulli, $\eta\,,\,\tau,\,\lambda$ siano le

$$(2)\quad b_{\eta\tau\lambda}^{(r)} = -\frac{1}{C}\left\{ \int_{\mathcal{C}} x_1^{\eta}\, x_2^{\tau}\, x_3^{\lambda}\, \mu\, F_r \ d\mathcal{C} + \int_{\Sigma} x_1^{\eta}\, x_2^{\tau}\, x_3^{\lambda}\, f_r \ d\Sigma \right\} \ ,$$

le coordinate iperastatiche della sollecitazione totale.[1]

---

[1] Facilmente generalizzabili in presenza di forze concentrate.

202

G. Grioli

Le equazioni cardinali della Meccanica si scrivono

$$b_{000}^{(r)} = 0 \quad , \quad (r = 1, 2, 3)$$

(3)

$$b_{100}^{(2)} - b_{010}^{(1)} = 0, \qquad b_{100}^{(3)} - b_{001}^{(1)} = 0, \qquad b_{010}^{(3)} - b_{001}^{(2)} = 0.$$

Per una qualunque funzione $f(P)$ integrabile in $\mathcal{C}$ si ponga

(4)
$$\bar{f} = \frac{1}{\mathcal{C}} \int_{\mathcal{C}} f \, d\mathcal{C}$$

Da (1) si trae facilmente

(5)
$$\eta \overline{X_{r1} \, x_1^{\eta-1} \, x_2^{\tau} \, x_3^{\lambda}} + \tau \overline{X_{r2} \, x_1^{\eta} \, x_2^{\tau-1} \, x_3^{\lambda}} + \lambda \overline{X_{r3} \, x_1^{\eta} \, x_2^{\tau} \, x_3^{\lambda-1}} = b_{\eta\tau\lambda}^{(r)},$$

$$(r = 1, 2, 3).$$

Le (5) generalizzano note relazioni di media di Signorini $\begin{bmatrix} 1 \end{bmatrix}$ che coincidono con le (5) per $n = \eta + \tau + \lambda = 1, 2$. Anzi, in tal caso le quantità a primo membro di (5) sono tutte determinate dalle (5) stesse. Invece, per $n > 2$ le equazioni (5) sono in numero inferiore a quello dei valori medi che contengono. Tuttavia, è facile riconoscere che per ogni $n > 2$ il sistema (5) permette di determinare quindici dei valori medi in esso contenuti. Precisamente, si ha

G. Grioli

$$(6) \begin{cases} n\; \overline{\overset{X}{\underset{r1}{}}\; \overset{n-1}{\underset{1}{x}}} = b^{(r)}_{n00} \;,\quad n\; \overline{\overset{X}{\underset{r2}{}}\; \overset{n\cdot1}{\underset{2}{x}}} = b^{(r)}_{0n0} \;,\quad n\; \overline{\overset{X}{\underset{r3}{}}\; \overset{n-1}{\underset{3}{x}}} = b^{(r)}_{00n} \;, \\[2em] n\; \overline{\overset{X}{\underset{11}{}}\; \overset{n+1}{\underset{1}{x}}\, \underset{2}{x}} = b^{(1)}_{n10} - \dfrac{b^{(2)}_{n+1\,0\,0}}{n+1} \;,\qquad \text{ecc.} \end{cases}$$

Se P_t è un qualunque polinomio nelle X_r , le quantità $\overline{X_{rs}P_t}$ risultano combinazioni lineari di soluzioni dei sistemi (5) corrispondenti ad opportuni valori di n . Naturalmente, tali quantità risultano determinate da tali sistemi se P_t è combinazione lineare di quei soli monomi che intervengono nelle (6) . In tal caso dirò che $\overline{X_{rs}P_t}$ è di classe M .

2. Limitazioni per lo stato tensionale

Per le componenti dello stress userò spesso notazioni ad un solo indice. secondo la convenzione

$$(7)\qquad X_r = X_{rr} \;,\qquad\qquad X_{r+3} = X_{r+1\,r+2} \;.$$

Dette a_t , $(t = 0, 1, \ldots\ldots, m)$ e β_s , $(s = 1, 2, \ldots\ldots, 6)$ delle costanti arbitrarie, risulta

$$(8)\qquad \oint \sum_{s=1}^{6} \beta_s X_s \sum_{t=0}^{m} a_t P_t \, dC = C \sum_{s=1}^{6}{}' \beta_s \sum_{t=0}^{m} a_t \overline{X_s P_t} \;,$$

da cui segue

G. Grioli

$$
(9) \qquad \left| \sum_{s=1}^{6} \beta_s X_s \right|_{max} > \frac{C \left| \sum_{s=1}^{6} \beta_s \sum_{t=0}^{m} a_t \overline{X_s P_t} \right|}{\int_C \left| \sum_{t=0}^{m} a_t P_t \right| dC}
$$

ove si è omesso il segno di uguaglianza che può capitare in casi veramente eccezionali e di facile individuazione.

Da (9) segue l'interessante limitazione

$$
(10) \qquad |X_r|_{max} > C \frac{\left| \sum_{t=0}^{m} a_t \overline{X_r P_t} \right|}{\int_C \left| \sum_{t=0}^{m} a_t P_t \right| dC}
$$

Le a_t possono determinarsi in modo che il secondo membro di (9) $\left[\text{e di } (10) \right]$ sia massimo ma ciò può riuscire laborioso. Una scelta conveniente delle a_t è

$$
(11) \qquad a_t = \sum_{s=1}^{6} \beta_s \frac{\overline{X_s P_t}}{\varrho_t^2}
$$

con

$$
(12) \qquad \varrho_t^2 = \frac{1}{C} \int_C P_t^2 \, dC .
$$

come risulterà chiaro nel § 3.

Identificando il polinomio $\sum_{s=1}^{6} \beta_s P_s$ successivamente con quello che espri-

G. Grioli

me lo sforzo normale, di taglio o - nel caso dei corpi elastici poco deformabili - con quello che esprime le caratteristiche di deformazione o le dilata-
zioni, gli scorrimenti, le variazioni di temperatura nel caso adiabatico, ecc.,
la (9) dà le corrispondenti limitazioni per i massimi moduli. Ad es., posto

$$(13) \qquad \Psi_\nu = \sum_{s=1}^{3} (\nu_s^2 X_s + 2\nu_{s+1\,s+2} X_{s+3}) = \sum_{s=1}^{3} \beta_s X_s \,,$$

lo sforzo normale $\phi_{\nu\eta}$ relativo alla direzione orientata caratterizzata dal
vettore unitario $\underline{\nu}$ soddisfa alla limitazione

$$(14) \qquad |\phi_{\nu\eta}|_{max} > C \; \frac{\left| \sum_{t=2}^{m} a_t \overline{\Psi_\nu} P_t \right|}{\int_{k} \left| \sum_{t=r}^{m} a_t P_t \right| d\zeta}$$

qualunque siano le costanti a_t

Si supponga che C sia un corpo cilindrico di sezione A , riferito
ad una terna centrale di cui l'asse x_1 sia parallelo alle generatrici. Si
assuma

$$(15) \qquad P_0 = 1 \,, \qquad P_1 = 0 \,, \qquad P_2 = x_2, \qquad P_3 = x_3, \qquad P_t = 0, \qquad se\ t > 3.$$

Fissata una sezione $x_1 = $ cost. e dette $R_i'(x_1)$ le componenti del risultante delle forze esterne agenti sulla parte di C compresa tra la sezione considerata e la base con $x_1 < 0$ e $M_i'(x_1)$ quelle del loro momento

G. Grioli

risultante rispetto al baricentro della sezione considerata, siano R_i, M_i i valori medi di $R'_i(x_1)$, $M'_i(x_1)$ al variare di x_1 tra le due basi di $\mathcal{C}$. Da (10) segue

$$(16) \qquad \left| X_{11} \right|_{max} > \frac{a_o R_1 - a_2 M_3 + a_3 M_2}{\displaystyle\int_A \left| a_o + a_1 x_1 + a_3 x_3 \right| dA} \;.$$

Supposto $R_1 \geqslant 0$ e detta r la retta del piano $x_1 = 0$ di equazione

$$(17) \qquad R_1 - \frac{M_3}{\rho_2^2} y_2 + \frac{M_2}{\rho_3^2} y_3 = 0 \;,$$

sia A_1 la parte di A ove il primo membro di (17) è positivo oppure l'intera A se r non taglia A. Inoltre, siano G_1 il baricentro di A, δ la sua distanza da r e $\underline{M}^*$ il vettore di componenti $0, \dfrac{M_2}{\rho_3^2}, \dfrac{M_3}{\rho_2^2}$.

Si può dimostrare che se le a_t sono espresse da (11) sussiste una delle due limitazioni :

$$(18) \qquad \begin{cases} \left| X_{11} \right|_{max} > \dfrac{R_1^2 + \underline{M} \cdot \underline{M}^*}{2\, A_1\, \underline{M}^* \delta - A R_1} \;, \\[3em] \left| X_{11} \right|_{max} > \dfrac{R_1^2 + \underline{M} \cdot \underline{M}^*}{A\, \underline{M}^* \delta} \;, \end{cases}$$

delle quali vale la prima se l'antipolo Q di r rispetto all'ellisse centra-

G. Grioli

le di   A   è interno al nocciolo centrale di   A   , la seconda se esso è esterno $\left[\,2\,\right]$.

Si riconosce che   r   e   Q   generalizzano i concetti di <u>asse neutro</u> e di <u>centro di pressione</u> nel classico problema di De Saint Venant della presoflessione.

- - - - - - - - - - - - - - - - -

Si supponga ora che $\mathcal{C}$ sia un prisma rettangolare la cui sezione abbia i lati di rispettive lunghezze b_2 e $b_3 \lessgtr b_2$ e inoltre risulti $R_1 = 0$. In tal caso, posto $a_0 = 0$, non è difficile determinare a_1 , a_2 in modo da rendere massimo il secondo membro di (16) . Ne segue il seguente risultato $\left[\,2\,\right]$.

Posto $q = \dfrac{b_3}{b_2} \leqslant 1$, nella sezione $x_1 = 0$ si considerino le parabole $\mathcal{P}_1$, $\mathcal{P}_2$ di rispettive equazioni

$$x_3^2 + \frac{2\,b_2\,q^2}{3}\,x_2 - \frac{b_3^2}{3} = 0 \ ,$$

(19)

$$x_2^2 + \frac{2\,b_3}{3\,q^2}\,x_3 - \frac{b_2}{3} = 0 \ ,$$

e la retta di equazione

(20) $$x_3 - q\,x_2 = 0.$$

Siano : B_1 la porzione di   A   delimitata dalla parabola $\mathcal{P}_1$, dalla retta di equazione   (20)   e dalle condizioni $x_2 \geqslant 0$, $x_3 \geqslant 0$; B_2 quella delimitata da $\mathcal{P}_2$, da $x_2 \geqslant 0$, $x_3 \geqslant 0$ e dalla retta di equazione (20) ;

B la porzione della sezione A costituita da $B_1 + B_2$ e dalle sue sim-
metriche rispetto agli assi x_2, x_3. Si ha la seguente condizione di sicu-
rezza : <u>condizione necessaria affinchè</u> $|X_{11}|_{max}$ <u>non superi una quantità po-
sitiva h^2 è che il punto di coordinate</u> $0, \dfrac{M_3}{Ah^2}, \dfrac{M_2}{Ah^2}$ <u>sia interno a</u>
B . Si riconosce facilmente che B è interna alla ellisse centrale della se-
zione A dal che si deduce che la condizione enunciata migliora un'analoga
condizione già data da Signorini $\begin{bmatrix}3\end{bmatrix}$.

3. <u>Altre limitazioni fondamentali per lo stress.</u>

Le relazioni integrali (5) possono sfruttarsi per dare concretezza anche
a certe nuove limitazioni per lo stress. Tali limitazioni possono riuscire
meno vantaggiose di quelle precedentemente stabilite ma sono di maggiore
portata, di più facile calcolo e permettono di stabilire, a volte, limitazioni
per le deformazioni.

Siano : q_{rs} i coefficienti costanti di una forma quadratica definita o almeno semidefinita positiva in sei variabili; γ_{rt} delle costanti arbitrarie;
$Q_o, \ldots, Q_m$, m+1 funzioni di x_1, x_2, x_3 definite in C e ivi tra loro ortogonali. Siano, inoltre,

$$(21) \qquad \mu_t^2 = \frac{1}{C} \int_C Q_t^2 \, dC$$

e

$$(22) \qquad \Psi = \sum_{r,s=1}^{6} \int_C q_{rs}\left(X_r - \sum_{t=0}^{m} \gamma_{rt} Q_t\right)\left(X_s - \sum_{t=0}^{m} \gamma_{st} Q_t\right) dC \;.$$

Da (21) segue

$$(23) \qquad \Psi = \sum_{r,s=1}^{6} \int_C q_{rs} X_r X_s \, dC + C \sum_{r,s=1}^{6} q_{rs} \sum_{t=0}^{m} \gamma_{st}\left(\mu_t^2 \gamma_{rt} - 2\overline{X_r Q_t}\right) > 0 \;,$$

che per

$$(24) \qquad \gamma_{rt} = \frac{\overline{X_r Q_t}}{\mu_t^2}$$

diviene

$$(25) \qquad \sum_{r,s=1}^{6} \int_C q_{rs} X_r X_s \, dC \geqslant C \sum_{t=0}^{m} \sum_{r,s=1}^{6} q_{rs} \frac{\overline{X_r Q_t}\,\overline{X_s Q_t}}{\mu_t^2}$$

Si riconosce facilmente $[4]$ che le γ_{rt} espresse da (24) minimizzano la ψ se la forma quadratica di cui i q_{rs} sono i coefficienti è proprio definita positiva. Anzi, si riconosce che nella (25) vale il segno di uguaglianza se e solo se risulta

$$(26) \qquad X_r = \sum_{t=0}^{m} \frac{\overline{X_r Q_t}}{\mu_t^2} Q_t \ .$$

Ne segue : <u>tra tutti gli stress corrispondenti ai medesimi valori medi $\overline{X_r Q_t}$, quello che minimizza in media ogni forma quadratica definita positiva nelle componenti dello stress è espresso da</u> (26).

La limitazione acquista interesse concreto ogni qualvolta gli $\overline{X_r Q_t}$ siano conoscibili, come accade, in particolare, se le funzioni Q_t coincidono con i polinomi P_t di cui nei paragrafi precedenti. In tal caso la (25) si scrive

$$(27) \qquad \sum_{r,s=1}^{6} \overline{q_{rs} X_r X_s} \geqslant \sum_{t=0}^{m} \sum_{r,s=1}^{6} q_{rs} \frac{\overline{X_r P_t}\,\overline{X_s P_t}}{\rho_t^2}$$

L'applicazione della disuguaglianza di Schwarz permette di dimostrare $[4]$ che ogniqualvolta le (9), (27) sono confrontabili in quanto sussiste l'uguaglianza

$$(28) \qquad \sqrt{\sum_{r,s=1}^{6} q_{rs} X_r X_s} = \left| \sum_{s=1}^{6} \beta_s X_s \right| ,$$

la limitazione (9) è più efficace della (27) per opportuna scelta delle a_t. Ciò capita, in particolare, quando si assumano per le a_t proprio i valori espressi da (11).

Da (27) segue immediatamente

$$(29) \qquad \left| X_r \right|_{max} \geqslant \sqrt{ \sum_{t=0}^{m} \frac{\overline{X_r P_t}^2}{\rho_t^2} }$$

con il secondo membro noto se le forze esterne sono note e gli $\overline{X_r P_t}$ di classe M .

Tra le limitazioni concrete che possono dedursi da (27) segnalo soltanto una condizione necessaria di plasticità. Precisamente, tenendo presente la nota condizione di von Mises, da (27) si deduce che condizione necessaria affinchè un corpo sia al disotto della soglia plastica è che risulti

$$(30) \qquad \sum_{t=0}^{m} \frac{1}{\rho_t^2} \sum_{s=1}^{3} \left[\left(\overline{X_{ss} P_t} - \overline{X_{s+1s+1} P_t} \right)^2 + b \, \overline{X_{ss+1} P_t^2} \right] < b^2$$

ove b^2 caratterizza il limite elastico.

La condizione (30) dà in effetti una condizione per le forze esterne.

4. <u>Limitazioni per gli spostamenti dei corpi elastici poco deformabili.</u>

In questo paragrafo supporrò che C sia la configurazione di equilibrio di un corpo elastico per cui vale la teoria lineare classica. La deformazione si penserà, pertanto, valutata a partire da una configurazione naturale vicinissima C^*. Le componenti dello stress X_{rs} possono pensarsi, com'è abituale nella teoria classica, quali dirette funzioni delle coordinate del generico punto P^* di C^*, nel senso che esse - che in una teoria con deformazioni finite hanno il significato di componenti euleriane dello stress - si devono ritenere coincidenti con le componenti lagrangiane, Y_{rs}, delle caratteristiche di tensione. In definitiva, in armonia con la teoria lineare, ai fini della valutazione degli inte-

G. Grioli

grali il campo d'integrazione si identifichrà con C^{*} , le coordinate di P
con le coordinate y_i , del corrispondente P^{*} in C^{*} , le X_{rs} con le
Y_{rs} .

Identificando i coefficienti q_{rs} con quelli, m_{rs} , della forma quadratica

$$(31) \qquad W = \frac{1}{2} \sum_{r,s=1}^{6} m_{rs} Y_r Y_s$$

che esprime la densità di energia potenziale elastica, il teorema di Clapey-
ron e la (27) permettono di dedurre per il lavoro delle forze esterne nello
spostamento $C^{*} \longrightarrow C$ e per l'energia elastica totale la limitazione

$$(32) \qquad \int L^{(e)} = 2 \int_{C^{*}} W\,dC^{*} \geqslant C^{*} \sum_{t=0}^{m} \frac{1}{\varrho_t^2} \sum_{r,s=1}^{6} m_{rs} \overline{Y_r P_t}\;\; \overline{Y_s P_t} \; ,$$

Varie applicazioni possono farsi della (32) . Ad es., nel caso di un corpo incastrato o appoggiato senza attrito su una parte del suo contorno e soggetto a forze esterne attive di direzione e verso invariabili, come nel caso
del peso, il lavoro delle reazioni vincolari è nullo e la (32) permette di dedurre per la componente u dello spostamento nella direzione della forza
la limitazione

$$(33) \qquad |u|_{max} \geqslant \frac{C^{*}}{R} \sum_{t=0}^{m} \frac{1}{\varrho_t^2} \sum_{r,s=1}^{6} m_{rs} \overline{Y_r P_t}\;\; \overline{Y_s P_t} ,$$

ove R indica il modulo del risultante della sollecitazione attiva. Interessanti limitazioni si deducono da (32) per elementi caratteristici della defor-

G. Grioli

mazione nel caso che la sollecitazione esterna attiva agisca solo su due por-
zioni σ', σ'' del contorno σ del corpo mentre sulla rimanente parte
non ci sono forze nè vincoli.

Se le forze agenti singolarmente sulle due parti σ', σ'' sono ridu-
cibili a due singole forze agenti una su σ', l'altra su σ'' o a due singole coppie, si trovano limitazioni per un certo allungamento o per una certa
rotazione che richiamano risultati noti della teoria di De Saint Venant e che
spesso mostrano come le deformazioni cui porta quella teoria sono più pic-
cole di quelle volute dalla teoria lineare esatta $[5], [6]$.

Qualche interessante limitazione può dedursi dalla (32) anche per le variazioni di volume delle singole cavità nel caso di corpi con cavità (involu-
cri) $[7]$.

Se si denota con τ la forma lineare nelle Y_r :

$$(34) \qquad \tau = \sum_{s=1}^{6} \beta_s Y_s$$

che esprime la variazione di temperatura di un corpo isotropo nel passag-
gio adiabatico da C^* a C , da (32) si deduce

$$|\tau|_{max} > \frac{L}{\left[3\left(\lambda + \dfrac{L^2}{c}\right) + 2\mu\right] c} \sqrt{\sum_{t=0}^{m} \frac{1}{\varrho_t^2} \, I_1(Y)P_t^2} \geqslant$$

$$\geqslant \frac{L}{\left[3\left(\lambda + \dfrac{L^2}{c}\right) + 2\mu\right] c} \sqrt{I_1^2 + \sum_{t=1}^{3} \cdot \frac{\overline{I_1 y_t}^2}{\varrho_t^2}} \quad ,$$

213

G. Grioli

ove $I_1 = Y_1 + Y_2 + Y_3$, λ e μ sono le costanti di Lamè , L il coefficiente di tensione termica e c dipende dal calore specifico a **configurazione** costante $\begin{bmatrix} 4 & , \text{pag.}\,86 \end{bmatrix}$.

OSSERVAZIONE Tutte le limitazioni sin qui stabilite presuppongono per la loro concreta valutazione la completa conoscenza delle forze esterne. In presenza di vincoli intervengono le reazioni vincolari generalmente incognite e quasi sempre non eliminabili dai secondi membri delle disuguaglianze stabilite.

Tuttavia in tale eventualità le disuguaglianze stabilite avranno i secondi membri noti se essi si minimizzano rispetto a tutti gli elementi indeterminati che essi contengono $\begin{bmatrix} \text{valori medi } \ \overline{Y_r P_t} \ \text{ e reazioni} \end{bmatrix}$, minimizzazione che va fatta tenendo conto non solo delle (5) ma anche della classe di sistemi di reazioni consentite dai vincoli.

G. Grioli

BIBLIOGRAFIA

[1] A. Signorini "Sopra alcune questioni di Statica dei sistemi continui" Ann. Scuola Norm. Sup. Pisa; Ser. II, 2, 3 - (1933).

[2] G. Grioli "Limitazioni per lo stato tensionale di un qualunque sistema continuo" Ann. Mat. Pura Appl. ser. IV, $\underline{39}$, 225-266 (1955).

[3] A. Signorini "Sopra un'estensione della teoria linearizzata dell'elasticità" Rend. Sem. Univ. Pol. Torino, $\underline{12}$, 83-93 (1952-53).

[4] G. Grioli "Mathematical Theory of Elastic Equilibrium (Recent Results)" Springer-Verlag. Berlin Gottingen Heidelberg (1962).

[5] G. Grioli "Sullo stato tensionale dei continui in equilibrio e sulle deformazioni nel caso elastico" Conferenze Sem. Mat. Univ. Bari 35-36 (1958).

[6] E. Bentsik " Sulle deformazioni elastiche dovute ad una sollecitazione riducibile a due coppie in equilibrio "Rend. Sem. Mat. Padova V. XXXIII; Pag. 297 (1963).

[7] G. Grioli "Sulle deformazioni elastiche di un involucro omogeneo soggetto a pressione o trazione" Rend. Sem. Mat. Univ. Padova $\underline{20}$, 278-285, (1951).

G. Grioli

PROBLEMI DI INTEGRAZIONE NELLA TEORIA
DELL'EQUILIBRIO ELASTICO

Lettura I

Sul teorema di Menabrea. Considerazioni sulla questione di esistenza

1. Considerazioni introduttive. Richiamo del teorema di Menabrea.
L'applicazione delle proprietà di media precedentemente indicate permette
di costruire un metodo d'integrazione del problema di equilibrio elastico valido non solo nel caso lineare ma addirittura in quello delle deformazioni finite.
Il metodo assume quali dirette incognite anzichè le componenti dello spostamento quelle dello stress le quali, del resto, quasi sempre sono gli elementi
di maggiore interesse concreto.

Nella prima lezione saranno svolte delle considerazioni sul teorema di
Menabrea che da un lato sono essenziali per la costruzione del metodo d'integrazione mentre dall'altro permettono di tradurre lo stesso teorema di esistenza della soluzione in quello dell'esistenza del minimo di un certo funzionale operante sullo stress e qui si noterà l'analogia della circostanza con
quanto ha già stabilito Signorini [12] operando sulle componenti di spostamento e dando luogo ad interessanti risultati di G. Fichera [3] a proposito
di un problema dallo stesso Signorini definito con la locuzione ''dalle ambi-
gue condizioni al contorno'' [1] .

Le prime due letture si riferiscono a corpi elastici poco deformabili per
i quali vale la teoria classica lineare. E' pertanto lecito, com'è abituale, con-
fondere lo stato attuale con quello noto di riferimento ai fini del calcolo degli

G. Grioli

integrali che s'incontrano e della valutazione delle forze esterne. L'energia elastica è espressa da una forma quadratica definita positiva con coefficienti indipendenti dalla temperatura il che è quanto dire che si prendono in considerazione trasformazioni isoterme o adiabatiche. Per semplicità, si considereranno deformazioni a partire da uno stato naturale - cioè, esente da stress - di corpi elastici anche anisotropi e - limitatamente alla prima lettura - inomogenei.

Si suppone l'assenza di vincoli interni, ma è noto $\left[2,\ \text{pag.}\,363...\right]$ che il caso dei solidi elastici incomprimibili si può far rientrare in quello dei solidi comprimibili semplicemente **particolarizzando** i valori dei coefficienti dell'espressione dell'energia elastica.

Al solito, denoto con C^* la configurazione di riferimento (stato naturale) di un solido elastico poco deformabile e con Σ^* la sua frontiera, supponendo che in corrispondenza alla parte Σ' di Σ^* siano note le forze superficiali mentre sull'altra parte, Σ'', vi siano dei vincoli per i quali si sa caratterizzare l'insieme Γ di tutti i possibili sistemi di reazione di cui essi sono capaci. Il sistema di Cauchy esplicitamente si scrive

$$
(1) \qquad
\begin{cases}
Y_{rs}^{'s} = k\, F_r^* \ , & (\text{in } C^*) \ , \\[2ex]
Y_{rs}\, N^s = f_{r'}^* \ , & (\text{su } \Sigma') \ ,
\end{cases}
$$

$$
(2) \qquad Y_{rs}\, N^s = \Phi_r \qquad\qquad (\text{su } \Sigma'') \ ,
$$

G. Grioli

con ϕ_r appartenente a Γ.

Sia ΔY_{rs} un'arbitraria variazione dello stress, soddisfacente al sistema

$$(3) \quad \begin{cases} (\Delta Y_{rs})^{'s} = 0 , & (\text{in } \overset{*}{C}) , \\[2mm] \Delta Y_{rs} N^s = 0 , & (\text{su } \Sigma') \end{cases}$$

e sia su Σ''

$$(4) \quad \Delta \phi_r = \Delta Y_{rs} N^s ,$$

La corrispondente variazione della densità di energia potenziale elastica - espressa nelle Y_{rs} - è[1]

$$(5) \quad \Delta W = \sum_{s=1}^{6} \frac{\partial W}{\partial Y_s} \Delta Y_r + W(\Delta Y_r).$$

Detto Y_{rs}^* lo stress effettivo o - in assenza di unicità - uno stress congruente verificante le equazioni di Cauchy e u_r^* il corrispondente spostamento, da (3), (4), (5) segue

$$(6) \quad \Delta \int_{C^*} W(Y_{rs}^*) \, dC^* - \int_{\Sigma''} \Delta \phi_r u^{*r} \, d\Sigma'' > 0 ,$$

---

(1) Si usano notazioni ad un solo indice analoghe a quelle usate nella Lettura I..

G. Grioli

valida per ogni scelta delle ΔY_{rs} non tutte nulle, verificanti le (3) e tali da dar luogo, in base a (2) , (4) , a reazioni $\phi_r + \Delta \phi_r$ appartenenti a Γ . La (6) esprime, com'è ben noto, il teorema di Menabrea e caratterizza lo (o uno) stress congruente.

2. Qualche conseguenza del teorema di Menabrea. Questione di esistenza.
Se u_r^* è assegnato su Σ'' il teorema di Menabrea caratterizza la soluzione come quella che rende minimo $\left[\text{ vedi } (\hat{6}) \right]$

$$(7) \qquad A = \int_{C^*} V'(Y_{rs}^*) \, dC^* - \int_{\Sigma'} \phi_r \, u^{*r} \, d\Sigma''$$

nella classe degli stress verificanti le (1). In tale caso è compreso quello dell'incastro su Σ'' , per $u_r^* = 0$. Interessante è il caso che su Σ'' siano assegnate le modalità di realizzazione dei vincoli in modo che riesca possibile definire l'insieme Γ e il tipo di spostamento da essi consentito, pur rimanendo incognito quello effettivamente realizzato. Trattasi, in effetti di problemi sostanzialmente non lineari a causa di disuguaglianze nelle condizioni al contorno e l'assumere come dirette incognite le componenti di spostamento può non riuscire conveniente e dar luogo a difficoltà che con vantaggio possono eliminarsi riferendosi direttamente alle caratteristiche dello stress, sulla base del sistema (1), (2) , associato al teorema di Menabrea. Così operando, trattasi di determinare, tra tutti gli stress verificanti le (1) , (2) , quello che minimizza A , lasciando ϕ_r in Γ . Si è così sicuri che lo stress trovato è congruente ma rimane legittimo il dubbio se lo spostamento u_r dedotto per integrazione di ben note espressioni nelle Y_{rs}^* e che contiene di arbitrario al più uno spostamento rigido, verifichi le condizioni geometrico-cinematiche volute dai vincoli.

G. Grioli

Se per il problema considerato negli spostamenti esiste un teorema di esistenza e di unicità il dubbio naturalmente cade, com'è evidente, sempre che sia stata caratterizzata in modo corretto la classe Γ .

Ciò accade nel caso di un appoggio liscio con un supporto non cedevole quando le forze esterne attive appartengono ad una certa classe, come di recente ha dimostrato G. Fichera $[3]$ studiando un interessante problema semilineare posto da A. Signorini $[1]$.

A me qui pare interessante osservare che il teorema di Menabrea, opportunamente inteso, permette di prescindere dal teorema di esistenza e di unicità nel problema sugli spostamenti nel senso che esso permette di tradurre la condizione di esistenza in quella di esistenza del minimo di un certo funzionale che opera direttamente sullo stress. Denominando <u>stress matematico</u> ogni soluzione delle (1) , si deve precisamente ritenere vero il teorema: <u>se</u> A <u>ammette un minimo nell'insieme degli stress matematici che danno luogo a reazioni appartenenti alla classe</u> Γ , <u>la soluzione del problema di equilibrio esiste</u>. Ciò nel senso che lo stress Y^{*}_{rs} minimizzante A nell'insieme degli stress matematici tenuto conto della classe Γ , non solo è (com'è ben noto) congruente ma dà luogo a spostamenti verificanti quasi ovunque le condizioni imposte dai vincoli. Il teorema di esistenza è pertanto tradotto in quello dell'esistenza del minimo di A in un'opportuna classe.

3. <u>Precisazione del teorema precedente. Dimostrazione in casi fisicamente interessanti</u>. I casi di maggiore interesse concreto sono indubbiamente quelli in cui è presente il vincolo d'incastro o di appoggio bilaterale o unilaterale (con supporto rigido o cedevole) e in tali casi darò la dimostrazione del teorema enunciato. Il primo di tali casi rientra, com'è stato già osserva-

G. Grioli

to, in quello di assegnato spostamento su $\sum{}''$. Sia $\bar{u}_r$ lo spostamento assegnato su $\sum{}''$ e si supponga che qualunque reazione sia eplicabile dal vincolo $\left[\int{}' \text{ contiene qualunque vettore}\right]$ come accade nel caso d'incastro. Il teorema del numero precedente diviene ora : <u>Se su $\sum{}''$ è assegnato lo spostamento u_r'' , la soluzione esiste se e solo se esiste il minimo di</u>

$$(8) \qquad A' = \int_{C^*} W(Y_{rs})\, dC^* - \int_{\Sigma''} \phi_r\, \bar{u}^r\, d\sum{}'' \quad ,$$

<u>nella classe di tutti gli stress matematici. Lo stress minimante coincide con quello effettivo.</u> La dimostrazione è immediata[2]. Basterà far vedere che allo stress Y_{rs}^* minimizzante A' nella classe di tutti i possibili stress matematici corrisponde uno spostamento u_r' coincidente (quasi ovunque) con $\bar{u}_r$ su $\sum{}''$. A tal fine si osservi che la condizione di minimo di A' si esplicita, in base a (5) e alle note relazioni tra stress e strain nella disuguaglianza

$$(9) \qquad \int_{C^*} W(\Delta Y_{rs}^*)\, dC^* - \int_{C^*} \sum_{r,s=1}^{3} \Delta Y_{rs}^*\, u_{r,s}'\, dC^* - \int_{\Sigma''} \Delta \phi_r^*\, \bar{u}^r\, d\sum{}'' > 0$$

che, per le (3) , diviene

$$(10) \qquad \int_{C^*} W(\Delta Y_{rs}^*)\, dC^* + \int_{\Sigma''} \Delta \phi_r^*\, (u'^r - \bar{u}^r)\, d\sum{}'' \geqslant 0,$$

valendo il segno di uguaglianza solo per $\Delta Y_{rs} = 0$.

 Fissata una sestupla $\Delta Y_{rs}^* = \zeta_{rs}$, sia $q_r = \zeta_{rs} N^{rs}$. Se ζ_{rs} verifica le (3) , anche $\lambda \zeta_{rs}$ con λ costante le verifica e la (10) si traduce

---

(2) La necessità è evidente.

G. Grioli

in

$$(11) \qquad \lambda^2 \int_{C^{*}} W(\xi_{rs})dC^{*} + \lambda \int_{\Sigma''} q_r(u'^r - \bar{u}^r)d\Sigma'' \geqslant 0.$$

Dato che Y^{*}_{rs} minimizza A' nella classe di tutti gli stress matematici, la (11), fissato ξ_{rs}, deve valere per ogni λ e qualunque sia il suo segno. Ciò implica l'annullarsi di $q_r(u'^r - \bar{u}^r)$ quasi ovunque su Σ'' e poichè la larga arbitrarietà di ξ_{rs} rende certo possibile avere nella (11) dei vettori q_r con direzioni non tutte complanari[3], si deduce che su Σ'' risulta quasi ovunque $u'^r - \bar{u}^r = 0$.

Nel caso di vincolo d'incastro è $\bar{u}_r = o$ e conseguentemente $u'_r = o$.

Si supponga ora che il vincolo presente su Σ'' sia un appoggio unilaterale liscio cedevole elasticamente o, in particolare, rigido. L'insieme Γ comprende tutti i vettori paralleli e concordi a N_r e il vettore nullo. Supporrò che lo spostamento corrispondente alla reazione esplicata dal vincolo sia esprimibile nella forma

$$(12) \qquad u_r = v_r \begin{cases} -\dfrac{\phi}{\mu} N_r & \text{se } \phi > 0, \\[2ex] +\varrho\, N_r & \text{se } \phi = 0, \end{cases}$$

---

(3) Si tratta in effetti di ammettere l'esistenza di soluzioni del sistema (1), (2) con ϕ_r fissato in modo opportuno.

G. Grioli

ove v_r è un arbitrario spostamento tangenziale e μ una costante da supporsi infinita nel caso di appoggio su supporto rigido.

Il teorema del n. 2 si esplicita nei seguenti termini : Condizione neces-saria e sufficiente affinchè esista la soluzione del problema di equilibrio è che esista il minimo di

$$(13) \qquad A'' = \int_{C^*} W(Y_{rs})dC^* + \frac{1}{2\mu} \int_{\Sigma''} \phi^2 \, d\Sigma'' \quad ,$$

nell'insieme di tutti gli stress matematici che lasciano ϕ_r nella classe Γ

La necessità è una conseguenza immediata di (6), (12) . Per dimostrare la sufficienza, si denoti con u'_r lo spostamento indotto dalle Y_{rs} minimizzanti e con ϕ^*_r la corrispondente reazione. Sia Σ''_1 la parte di Σ'' ove risulta $\phi^*_r \neq 0$, Σ''_2 quella ov'è $\phi^*_r = 0$. La condizione di minimo di A'' , in base a (5) , si esplicita nella disuguaglianza

$$(14) \qquad \int_{C^*} W(\Delta Y^*_{rs})dC^* + \frac{1}{2\mu} \int_{\Sigma''} (\Delta \phi^*)^2 d\Sigma'' +$$

$$+ \int_{\Sigma''_1} \Delta \phi^*_r (u'^r + \frac{\phi N^r}{\mu}) d\Sigma''_1 + \int_{\Sigma''_2} \Delta \phi^*_r \, u'^r d\Sigma''_2 \geqslant 0 \quad ,$$

da ritenersi verificata per ogni scelta ΔY^*_{rs} soddisfacente alle (3) e alla condizione che il vettore $\phi^*_r + \Delta \phi^*_r$ appartenga a Γ . Precisamente, dato che considereremo i vettori $\Delta \phi^*_r$ piccoli di fronte a ϕ^*_r , le ΔY^*_{rs} dovranno soddisfare alle (3) e inoltre dar luogo a vettori $\Delta \phi^*_r$ paralleli ad N^*_r , con verso arbitrario su Σ''_1 , concordi ad N^*_r su Σ''_2 .

G. Grioli

Per una prefissata scelta $\Delta Y^{*}_{rs} = \zeta_{rs}$ tra le possibili, sia q_r il vettore $\Delta \phi^{*}_{r}$ corrispondente. Lo supporrò concorde ad N^{*}_{r} .Detta λ una costante, anche $\lambda \zeta_{rs}$ e λq_r soddisfano alle condizioni richieste, purchè sia $\lambda \geqslant 0$ se q_r non è nullo su tutta $\sum''_2$ .La (14) diviene

$$(15) \quad \lambda^2 \int_{C^{*}} W(\zeta_{rs})dC^{*} + \lambda \int_{\sum''_2} u'_r \, q^{r} \, d\sum'''_2 + \lambda \int_{\sum''_1} (u'_r + \frac{\phi}{\mu} N_r) \, q^{r} d\sum''_1 \geqslant 0$$

e sussiste, per ipotesi, per ogni scelta delle $\lambda \zeta_{rs}$ tra quelle consentite. Si assumano le ζ_{rs} in modo che sia$^{(4)}$ $q_r \neq 0$ su tutta $\sum''_2$ .Di conseguenza λ può essere positivo o negativo e la (15) implica che u'_r soddisfi alla (12) quasi ovunque su tutta $\sum''_1$ dato che q_r è parallelo ad N^{*}_{r} . Pertanto in (15) manca il terzo termine. Supposto invece q_r non nullo su $\sum''_2$ e concorde ad N^{*}_{r} ,si deve ritenere $\lambda \geqslant 0$. Supponendo proprio $\lambda > 0$,la (15) implica proprio $u'_r N^{*r} \geqslant 0$ che rispetta il vincolo su $\sum''_2$ ed esprime distacco dall'appoggio nel caso di effettiva disuguaglianza .

<u>Osservazione</u> In quanto si è precedentemente esposto si constata che in effetti dall'esistenza del minimo di A nei vari casi considerati segue che lo spostamento u'_r indotto dallo stress minimizzante verifica su $\sum''$ certe condizioni integrali da cui segue che le condizioni imposte dal vincolo sono verificate su $\sum''$ a meno di un insieme di misura nulla. Si deve osservare che il soddisfare in media quelle condizioni è in armonia con una visione integrale della Meccanica dei continui - in particolare della teoria matematica dell'Elasticità - che, forse, solo ha senso fisico. Ciò d'altronde è stato già rimarcato da vari Autori in modo più o meno esplicito.

---

(4) Vedi nota (3) .

G. Grioli

BIBLIOGRAFIA

[1] A. Signorini ''Questioni di elasticità non linearizzata e semilinearizzata''
Rend. di Mat. Roma, vol. 18 ; 95-139 (1959)

[2] A. Signorini ''Trasformazioni termoelastiche finite - Solidi vincolati''
Ann. Mat. pura ed applicata; S. IV T. LI, 329-372 (1960)

[3] G. Fichera Presentato per la stampa in ''Atti dell'Accademia Nazionale
dei Lincei'' (Memorie). Roma (Italia).

G. Grioli

Lettura II

Integrazione del problema fondamentale dell'Elastostatica lineare.

1. **Premesse analitiche.** Sia $\left\{W_t\right\}$ la successione di tutti i possibili monomi formati con le coordinate y_1, y_2, y_3, ordinati per grado non decrescente. Sia $\left\{P_t\right\}$ la successione di polinomi costruita aggiungendo a W_t quella combinazione lineare di W_o, W_1, ..., W_{t-1} che rende P_t ortogonale a W_o, W_1, ..., W_{t-1}, nella regione C^* occupata dal corpo nel suo stato di riferimento. La successione $\left\{P_t\right\}$ è di polinomi ortogonali in C^* ed è costruibile in base ad un noto teorema di Gram. La supporrò inoltre, com'è certo possibile, normalizzata.

E' ben nota una proprietà di completezza di $\left\{P_t\right\}$. In base ad essa, condizione necessaria e sufficiente affinchè due funzioni ψ e φ definite la prima in C^*, la seconda sulla sua frontiera Σ^* siano quasi ovunque nulle rispettivamente in C^* e su Σ^* è che siano soddisfatte le infinite equazioni integrali [1]

$$(2.1) \qquad \int_{C^*} \psi\, P_t dC^* + \int_{\Sigma^*} \varphi\, P_t d\Sigma^* = 0, \qquad (t = 0, 1, \ldots),$$

Da tale proprietà di completezza segue che ogni funzione di quadrato sommabile in C^* è rappresentabile in serie dei P_t con convergenza in media. In particolare, ogni soluzione delle equazioni di equilibrio di Cauchy, che nelle notazioni ora adoperate si scrivono

G. Grioli

$$(2.2) \quad \begin{cases} Y_{rs}'^{\,s} = k^* F_r^* & \text{(in } C^*\text{)}, \\[2em] Y_{rs} N^s = f_r^* \, , & \text{(su } \Sigma^*\text{)}, \end{cases}$$

è esprimibile mediante una serie dei P_t . In effetti, la proprietà di completezza di $\left\{ W_t \right\}$ (identica a quella di $\left\{ P_t \right\}$) assicura che il sistema (2.2) è equivalente al sistema delle infinite equazioni integrali[5]

$$(2.3) \quad \eta \, \overline{Y_{r1} y_1^{\eta-1} y_2^{\tau} y_3^{\lambda}} \; + \; \tau \, \overline{Y_{r2} y_1^{\eta} y_2^{\tau-1} y_3^{\lambda}} \; + \; \lambda \, \overline{Y_{r3} y_1^{\eta} y_2^{\tau} y_3^{\lambda-1}} = b_{\eta\tau\lambda}^{(r)} \, ,$$

$$(r = 1, 2, 3),$$

nel senso che ogni soluzione delle (2.2) di quadrato sommabile in C^* soddisfa alle (2.3) , mentre ogni soluzione delle (2.3) per la quale gli $\overline{Y_r P_t}$ - che si costruiscono quali combinazioni lineari delle indeterminate del sistema (2.3) - siano tali che le serie

$$(2.4) \quad Y_r = \sum_{t=0}^{\infty} \overline{Y_r P_t} \cdot P_t$$

risultino convergenti in media, fornisce attraverso le (2.4) uno stress soddisfacente alle (2.2),

---

(5) Per le $b_{\eta\tau\lambda}^{(r)}$ valgono definizioni analoghe a quelle della Lettura I.

G. Grioli

2. <u>Appossimazioni polinomiali dello stress.</u> Si consideri la successione dei soli $m+1$ polinomi $P_o, P_1, \ldots\ldots, P_m$ e delle equazioni (2.3) solo quelle corrispondenti a tutti e soli i monomi W_t contenuti nei polinomi presi in esame. In corrispondenza ad una qualunque soluzione di tale sistema - che chiamerò sistema S_m - si costruisca la successione $\left\{ q_{rt}^{(m)} \right\}$, $r = 1, 2\ldots\ldots, 6; \ t = 0, 1\ldots\ldots, m)$ dei valori medi $\overline{Y_r P_t}$.

Detta al solito,

$$(2.5) \qquad W = \frac{1}{2} \sum_{r, s=1}^{6} m_{rs} Y_r Y_s$$

la densità di energia potenziale elastica, si ponga

$$(2.6) \qquad V_m = \frac{C^*}{2} \sum_{r, s=1}^{6} m_{rs} \sum_{t=0}^{m} q_{rt}^{(m)} q_{st}^{(m)}$$

e, inoltre,

$$(2.7) \qquad Y_r^{(m)} = \sum_{t=0}^{m} q_{rt}^{(m)} P_t .$$

Detta W_m ciò che diviene W quando si identificano le Y_r con le $Y_r^{(m)}$, si ha evidentemente

$$(2.8) \qquad V_m = \frac{1}{2} \int_{C^*} W_m \, dC^*$$

da cui si vede che V_m rappresenta l'energia totale corrispondente allo stress espresso da $Y_r^{(m)}$.

Al variare della considerata soluzione del sistema S_m la successione $\left\{ q_{rt}^{(m)} \right\}$ varia in un insieme che dirò I_m. Sia $\left\{ q_{rt}^{(m)*} \right\}$ la successione

231

G. Grioli

di I_m che minimizza V_m e siano $Y_r^{*(m)}$ e V_m^* le corrispondenti determinazioni di $Y_r^{(m)}$ e V_m . La sestupla $Y_r^{*(m)}$ rappresenta uno stress congruente.

Per dimostrarlo si denoti con $(k\,\overset{*}{\underset{m}{F}}\,,\,\overset{*}{\underset{m}{f}})$ la sollecitazione esterna corrispondente allo stress polinomiale $Y_r^{*(m)}$ in base alle (2.2) . Ogni soluzione delle (2.2) corrispondente a tale sollecitazione esterna è esprimibile nella forma

$$(2.9) \qquad Y_r = Y_r^{*(m)} + Y_r'$$

al variare di Y_r' nella classe di tutte le soluzioni del sistema omogeneo associato alle (2.2) . Stabilito per Y_r' uno sviluppo del tipo (2.4) , si ponga

$$(2.10) \qquad Y_r' = Y_r'^{(m)} + Y_r''\ ,$$

ove $Y_r'^{(m)}$ è il gruppo dei primi $m+1$ termini dello sviluppo, sicchè Y_r'' è ortogonale a $Y_r^{*(m)}$ e a Y_r' . L'energia totale corrispondente a (2.10) è

$$(2.11) \qquad V = \frac{C}{2}\left\{ \sum_{t=0}^{m} \sum_{r,s=1}^{6} m_{rs}\,(q_{rt}^{*(m)} + q_{rt}'^{(m)})(q_{st}^{*(m)} + q_{st}'^{(m)}) \; + \right.$$

$$\left. + \sum_{t=m+1}^{\infty} \sum_{r,s}^{6} m_{rs}\, q_{rt}''\, q_{st}'' \right\}$$

ove $q_{rt}'^{(m)}$, q_{rt}'' si riferiscono a $Y_r'^{(m)}$, Y_r'' , rispettivamente[6]. Poichè i due termini espressi dalle sommatorie in (2.11) sono singolarmente defini-

---

[6] Si suppone, cioè, $q_{rt}'^{(m)} = \overline{Y_r'^{(m)}P_t}$, $q_{rt}'' = \overline{Y_r''P_t}$.

G. Grioli

ti positivi e le $q'^{(m)}_{rt}$, q''_{rt} sono vincolate a soddisfare ad equazioni omo-

genee e pertanto possono assumere valori tutti nulli, si riconosce che il mi-

nimo di V si consegue quando le $q'^{(m)}_{rt}$, q''_{rt} sono proprio tutte nulle.

Ciò significa che lo stress $Y^{(m)}_r$ minimizza l'energia elastica totale nella

classe di tutte le soluzioni di (2.2) corrispondenti ad una speciale sollecitazio-

ne esterna e pertanto è congruente, in base al teorema di Menabrea, c.d.d..

E' ben naturale, dopo quanto si è detto, definire la sestupla

$$(2.12) \qquad Y^{(m)}_r = \sum_{t=0}^{m} q^{(m)}_{rt} P_t \ , \qquad\qquad (r = 1, 2, \ldots 6),$$

quale <u>approssimazione polinomiale di ordine</u> m <u>dello stress effettivo.</u>

Speciali cautele vanno osservate se su una parte $\sum''$ di $\sum^*$ sono pre-

senti dei vincoli. Per brevità mi riferirò al caso dell'incastro e dell'appoggio

liscio unilaterale con supporto rigido. In tali due casi, come risulta dalla let-

tura precedente, lo stress effettivo è proprio quello che minimizza l'energia

potenziale elastica - espressa nelle Y_r - nella classe di tutti gli stress ma-

tematici che danno luogo su $\sum''$ a reazioni appartenenti all'insieme Γ .

Si ponga

$$(2.14) \qquad b^{(r)}_{\eta\tau\lambda} = e^{(r)}_{\eta\tau\lambda} + c^{(r)}_{\eta\tau\lambda} \ ,$$

con

$$(2.15) \qquad c^{(r)}_{\eta\tau\lambda} = - \frac{1}{C} \int_{\sum''} \phi_r \, y_1^\eta \, y_2^\tau \, y_3^\lambda \, d\sum'' \ ,$$

mentre le $e^{(r)}_{\eta\tau\lambda}$ sono costruite in base alle (2) della lettura I in modo

evidente e con riferimento alla parte $\sum'$ di $\sum^*$, ove si suppongono assegna-

te le forze esterne.

G. Grioli

Il minimo di V_m va cercato al variare delle $q_{rt}^{(m)}$ nell'insieme I_m e delle $c_{\eta\tau\lambda}^{(r)}$ nell'insieme i_m definito in base a (2.15) al variare di ϕ_r in Γ e tenuto conto delle equazioni cardinali della Statica.

Ad es., se il vincolo è quello d'incastro la classe Γ comprende tutti i possibili vettori e le $c_{\eta\tau\lambda}^{(r)}$ sono tutte libere ad eccezione di quelle per le quali $\eta + \tau + \lambda = 0, 1$. Per esse risulta

$$(2.16) \qquad \begin{cases} c_{000}^{(r)} = - e_{000}^{(r)} \\[2em] c_{100}^{(2)} - c_{010}^{(1)} = e_{010}^{(1)} - e_{100}^{(2)} \ , \quad \text{ecc.} \end{cases}$$

Invece nel caso di un appoggio unilaterale rigido e liscio l'insieme Γ comprende tutti e soli i vettori ϕ_r soddisfacenti alla condizione

$$(2.17) \qquad \phi_r = \phi\, N_r \ ,$$

con $\phi \geqslant 0$. Il metodo presenta delle complcazioni se non ci si riferisce ad un opportuno sistema di coordinate curvilinee, tranne nel caso d'interesse concreto di un appoggio piano - o, anche formato di piu parti piane -. Se unico, si può fare in modo che l'appoggio piano appartenga al piano $y_3 = 0$ con l'asse y_3 concorde ad $\underline{N}$. In tale ipotesi valgono le equazioni di condizione

$$(2.18) \qquad \begin{aligned} & c_{000}^{(3)} = - e_{000}^{(3)} \ , \\[1.5em] & c_{100}^{(3)} = e_{001}^{(1)} - e_{100}^{(3)}, \qquad c_{010}^{(3)} = e_{001}^{(2)} - e_{010}^{(3)} \ , \end{aligned}$$

G. Grioli

mentre le altre $c_{\eta\tau\lambda}^{(r)}$ si devono ritenere nulle se $r = 1, 2$, libere se $r = 3$.

In generale, se lo stress $Y_{rs}^{*(m)}$ dà luogo, come può accadere, ad un vettore ϕ_r che su una parte Σ_ℓ' di Σ'' non soddisfa alla (2.17) in quanto risulti $\phi < 0$, si deve ritenere che su Σ_ℓ' ci sia distacco dal supporto di appoggio. In tale eventualità il procedimento di minimizzazione va ripetuto supponendo $\phi_r = 0$ su Σ_ℓ'' . Le $c_{\eta\tau\lambda}^{(r)}$ vanno allora definite ancora secondo le (2.15) ma sostitutendo al campo d'integrazione Σ'' l'altro, Σ_ℓ'' .

3. <u>Sull'integrazione del problema d'equilibrio.</u> Uno dei motivi per cui le $Y_r^{*(m)}$ espresse da (2.13) costituiscono una effettiva approssimazione dello stress reale è che sussiste il teorema: <u>detto V_m^* il minimo di V_m in I_m e, per ogni m ,detti $q_{rt}^{*(m)}$ i valori delle $q_{rt}^{(m)}$ minimizzanti, condizione necessaria e sufficiente affinchè la sestupla $Y_r^{*(m)}$ definita da (2.13) sia convergente in media verso la soluzione Y_r^* del problema di equilibrio è che la successione $\left\{V_m\right\}$ sia convergente.</u>

La necessità della condizione segue dal fatto che se il sistema delle equazioni di Cauchy ammette soluzioni di quadrato sommabile in C^* esse sono esprimibili, in base a (2.4), (2.7), nella forma

$$(2.19) \qquad\qquad Y_r = \lim_{m \to \infty} Y_r^{(m)}$$

e se $\left\{Y_r^{(m)}\right\}$ converge la successione $\left\{V_m\right\}$ al divergere di m converge anch'essa, per una nota proprietà del prodotto integrale. Del resto, per convincersene basta osservare che se $\left\{Y_r^{(m)}\right\}$ converge ad Y_r , si ha, per ogni ε positivo e a partire da un certo intero n

235

G. Grioli

$$(2.20) \qquad \left| Y_r - \sum_{t=0}^{m} q_{rt}^{(m)} P_t \right| < \varepsilon .$$

Segue

$$(2.21) \quad V - V_n = \frac{1}{2} \int_{C^*} \sum_{r,\,s=1}^{6} m_{rs} \left(Y_r - \sum_{t=0}^{m} q_{rt}^{(m)} P_t \right) \left(Y_s - \sum_{t=0}^{m} q_{st}^{(m)} P_t \right) dC^*$$

$$< \frac{C^*}{2} \, \varepsilon^2 \, \sum_{r,\,s=1}^{6} \left| m_{rs} \right| ,$$

che assicura la convergenza di $\left\{ V_m \right\}$ a V .

La convergenza di $\left\{ V_m \right\}$ assicura quella di $\left\{ V_m^* \right\}$ che è non decrescente.

Per dimostrare la sufficienza della condizione del teorema enunciato, supposta convergente la successione $\left\{ V_m^* \right\}$ al valore $V^* > 0$, si ponga

$$(2.22) \qquad V_{mt}^* = \frac{1}{2} \sum_{r,\,s=1}^{6} m_{rs} \, q_{rt}^{*\,(m)} \, q_{st}^{*\,(m)}$$

e, inoltre,

$$(2.23) \qquad \tau_t^{(m)} = \sqrt{ \sum_{r=1}^{6} q_{rt}^{*\,(m)\,2} } \qquad , \qquad q_{rt}^{*\,(m)} = \tau_t^{(m)} z_r$$

Da (2.23) segue

$$(2.24) \qquad \sum_{r=1}^{6} z_r^2 = 1 .$$

Per cose ben note, la forma quadratica $\displaystyle\sum_{r,\,s=1}^{6} m_{rs} \, z_r \, z_s$ ammette, sot-

G. Grioli

to la condizione (2.24) , un minimo positivo 2γ . Di conseguenza, in base a (2.22), (2.23), si ha

$$(2.25) \qquad V_{mt}^{*} > \tau_{t}^{(m)^2}\gamma \ .$$

Da (2.23) , (2.25) si deduce

$$(2.26) \qquad q_{rt}^{*(m)^2} < \tau_{t}^{(m)^2} < \frac{V_{mt}^{*}}{\gamma}$$

e quindi

$$(2.27) \qquad \lim_{m \to \infty} \sum_{t=0}^{m} q_{rt}^{*(m)^2} < \frac{1}{\gamma} \lim_{m \to \infty} \sum_{t=0}^{m} V_{mt}^{*} = \frac{1}{\gamma} \frac{V^{*}}{C} \ .$$

Posto, formalmente,

$$(2.28) \qquad q_{rt}^{*} = \lim_{m \to \infty} q_{rt}^{*(m)} \ ,$$

la (2.27) si scrive

$$(2.29) \qquad \sum_{t=0}^{\infty} q_{rt}^{*2} < \frac{V^{*}}{\gamma C^{*}}$$

la quale assicura che comunque si scelga un ε positivo arbitrariamente piccolo esiste un intero positivo n tale che per ogni intero positivo p si ha

$$(2.30) \qquad \sum_{t=n}^{n+p} q_{rt}^{*2} < \varepsilon \ .$$

G. Grioli

La (2.30) può scriversi

$$(2.31) \qquad \int_{C^*}\left[\sum_{t=n}^{n+P} q^*_{rt}\, P_t\right]^2 dC^* < \varepsilon .$$

La (2.31) , in base ad un noto criterio di Cauchy, assicura la convergenza in media di

$$(2.32) \qquad \sum_{t=0}^{\infty} q^*_{rt} P_t = \lim_{m\to\infty} \sum_{t=0}^{m} q^{*(m)}_{rt}\, P_t = \lim_{m\to\infty} Y^{*(m)}_r .$$

Posto

$$(2.32) \qquad Y^*_r = \lim_{m\to\infty} Y^{*(m)}_r ,$$

dall'essere, per ogni m , $V^*_m < V_m$, segue

$$(2.33) \qquad V^* = \lim_{m\to\infty} V^*_m < \lim_{m\to\infty} V_m = V .$$

Poichè V^* non può differire dall'energia totale elastica corrispondente a Y^*_r nè V da quella corrispondente alla generica sestupla Y_r di quadrato sommabile, la (2.33) mostra che la sestupla Y^*_r definita da (2.32) soddisfa al teorema di Menabrea nella classe di tutti gli stress matematici di quadrato sommabile che lasciano ϕ_r nella classe Γ e quindi congruente e, per quanto si è dimostrato nella lettura precedente, rispetta i vincoli, c.d.d..

G. Grioli

Osservazione. Da quanto sopra si deduce : condizione necessaria e sufficiente affinchè il problema fondamentale dell'Elastostatica lineare con forze assegnate su una parte della fromtiera e con vincolo d'incastro o d'appoggio liscio con supporto rigido sulla rimanente parte.abbia una ed una sola soluzione di quadrato sommabile è che il sistema fondamentale di Cauchy ammetta soluzioni di quadrato sommabile e che la successione $\left\{V_m^*\right\}$ sia convergente.

BIBLIOGRAFIA

[1] L. Amerio "Sul calcolo delle soluzioni dei problemi al contorno per le equazioni lineari del secondo ordine di tipo ellittico" Am. J. Math. LXIX, $\underline{3}$, 447 (1947)

G. Grioli

Lettura III

Sull'integrazione del problema fondamentale dell'Elastostatica nel caso delle deformazioni finite.

1. **Premessa generale.** Siano al solito C e C^* le configurazioni attuali e di riferimento, quest'ultima supposta di equilibrio naturale, $\sum$ e $\sum^*$ le loro frontiere, P e P^* punti corrispondenti di $C + \sum$ e $C^* + \sum^*$ e x_r, y_r le loro coordinate rispetto alla medesima terna trirettangola levogira.

Sia

$$(3.1) \qquad D \equiv \left\| x_{r,1} \right\| , \qquad\qquad x_{r,1} = \frac{\partial x_r}{\partial y_1}$$

e

$$(3.2) \qquad X_{rs} = \frac{1}{D} \, Y_{lm} \, x_{,r}^{\,l} \, x_{,s}^{\,m}$$

Le X_{rs} sono, com'è ben noto, le componenti euleriane dello stress, le Y_{rs} quelle lagrangiane. Il sistema fondamentale dell'Elastostatica si scrive $\left[\underline{1}, \text{pag.}\,34\right]$.

$$(3.3) \qquad (Y^{lm} \, x_{r,l})_{,m} = \theta k^* F_r^* , \qquad\qquad (\text{in } C^*) ,$$

$$(3.4) \qquad Y^{lm} \, x_{r,1} \, N_m^* = \theta f_r^* , \qquad\qquad (\text{su } \sum^*) ,$$

G. Grioli

ove $\theta k \overset{*}{F}_{r}$ è la forza specifica di massa riferita allo stato $\overset{*}{C}$ e $\overset{*}{f}_{r}$ la densità della forza superficiale esterna riferita a $\overset{*}{\sum}$. Le componenti dello spostamento $\overset{*}{C} \to C$ sono espresse da

$$(3.5) \qquad u_r = x_r - y_r \quad ,$$

mentre le caratteristiche di deformazione[7], ε_{rs} sono, notoriamente,

$$(3.6) \qquad \varepsilon_{rs} = \frac{1}{2} (u_{r,s} + u_{s,r} + u_{i,r} u^{i}_{,s}) \quad .$$

Limitandosi alle trasformazioni isoterme o adiabatiche, alle equazioni precedenti vanno associate le relazioni costitutive

$$(3.7) \qquad Y_r = - \frac{\partial W(\varepsilon)}{\partial \varepsilon_r} \quad , \qquad\qquad (r = 1, 2, \ldots, 6),$$

ove W è la densità di energia potenziale elastica[8]. Sarà bene tenere presente che, essendo $\overset{*}{C}$ stato naturale, per $\theta = 0$ lo stress è nullo :

$$(3.8) \qquad Y_{rs}^{(\theta = 0)} = 0 \quad , \qquad\qquad (\text{in } \overset{*}{C}) \quad .$$

Il problema analitico connesso alle equazioni (3.3), (3.4), (3.5), (3.6), (3.7) è notevolmente complesso. Mancano in generale teoremi di unicità, esistenza e sviluppabilità in serie di potenze del parametro θ della soluzio-

---

(7) A rigore le caratteristiche di deformazione sono le ε_{rr}, $2\varepsilon_{rs}$, $(r \neq s)$.

(8) Si usano per le ε_{rs} notazioni ad un solo indice analoghe a quelle già usate per le X_{rs} [vedi le (7) in Lettura I].

G. Grioli

ne, tranne nel caso in cui il vettore $\underline{f}^*$ sia noto sull'intera $\sum^*$.

In tal caso, sotto varie ipotesi concernenti i vettori $\underline{K}^*\underline{F}^*, \underline{f}^*$ la frontiera $\sum^*$ e per $|\theta|$ sufficientemente piccolo, teoremi siffatti sono stati stabiliti da F. Stoppelli $\begin{bmatrix} 1, 2, 3, 4, 5, 6 \end{bmatrix}$ sia nel caso in cui non vi siano assi di equilibrio come pure in quello (eccezionale) in cui ve ne siano. Qui non posso che rinviare ai lavori originali, limitandomi solo a ricordare che, almeno nel caso di forze superficiali ovunque note su $\sum^*$ e per $|\theta|$ sufficientemente piccolo si deve ritenere, salvo casi eccezionali (che possono presentarsi solo quando ci sono assi di equilibrio) che per larghi tipi di sollecitazioni esterne e di forma di frontiera la soluzione esiste ed è unica.

Nel caso, invece, di spostamenti assegnati su $\sum^*$ e nel problema misto mancano teorema generali del tipo di quelli di Stoppelli.

Lo scopo di quanto seguirà è di mostrare come il metodo d'integrazione precedentemente esposto sia adattabile al caso delle deformazioni finite, supposta esistente la soluzione. La validità del metodo abbraccia, pertanto, il caso di forze $\underline{f}^*$ ovunque note su $\sum^*$ e, formalmente, anche gli altri due (problema di assegnati spostamenti su $\sum^*$ e problema misto) e poichè l'intuizione fisica fa presumere l'esistenza della soluzione anche in tali casi e la possibilità di stabilire teoremi di esistenza e di unicità analoghi a quelli di Stoppelli, si è indotti a ritenere che la validità del metodo possa essere più che formale, almeno nel futuro, anche in riguardo agli altri due casi sopra menzionati.

2. <u>Esposizione del metodo d'integrazione.</u> La sviluppabilità in serie di potenze della soluzione - se c'è - del problema fondamentale si traduce nell'esistenza degli sviluppi

G. Grioli

$$(3.9) \qquad u_r = \sum_{i=1}^{\infty} u_r^{(i)} \frac{\theta^i}{i!} \, , \qquad\qquad \varepsilon_r = \sum_{i=1}^{\infty} \varepsilon_r^{(i)} \frac{\theta^i}{i!} \, ,$$

$$(3.10) \qquad Y_r = \sum_{i=1}^{\infty} Y_r^{(i)} \frac{\theta^i}{i!} \, ,$$

da ritenersi validi per $|\theta|$ sufficientemente piccolo[9].

Poichè W non dipende esplicitamente da θ, si ha

$$(3.11) \qquad \left(\frac{\partial W}{\partial \varepsilon_r}\right)^{(1)} = \sum_{s=1}^{6} \left(\frac{\partial^2 W}{\partial \varepsilon_r \partial \varepsilon_s}\right)_{\varepsilon=0} \varepsilon_s^{(1)} = \frac{\partial W^{(2)}}{\partial \varepsilon_r^{(1)}} \, ,$$

con

$$(3.12) \qquad \varepsilon_s^{(1)} = \frac{1}{2}\left(u_{r,s}^{(1)} + u_{s,r}^{(1)}\right)$$

e

$$(3.13) \qquad W^{(2)}(\varepsilon^{(1)}) = \frac{1}{2} \sum_{r,s=1}^{6} \left(\frac{\partial^2 W}{\partial \varepsilon_r \partial \varepsilon_s}\right)_{\varepsilon=0} \varepsilon_r^{(1)} \varepsilon_s^{(1)} = \frac{1}{2} \sum_{r,s=1}^{6} M_{rs} \varepsilon_r^{(1)} \varepsilon_s^{(1)}$$

In generale si ha, anzi,

$$(3.14) \qquad \left(\frac{\partial W}{\partial \varepsilon_r}\right)^{(n)} = \sum_{s=1}^{6} \left(\frac{\partial^2 W}{\partial \varepsilon_r \partial \varepsilon_s}\right)_{\varepsilon=0} \varepsilon_s^{(n)} + \text{termini che dipendono solo da}$$

$$\varepsilon_1^{(1)} , \ldots , \varepsilon_1^{(n-1)} \, .$$

---

(9) S'intende $g^{(n)} = \left(\dfrac{d g}{d\vartheta^n}\right)_{\vartheta=0}$

G. Grioli

La (3.14) può scriversi

$$(3.15) \qquad \left(\frac{\partial W}{\partial \mathcal{E}_r}\right)^{(n)} = \sum_{s=1}^{6} M_{rs} \, \mathcal{E}_s^{(n)} + \text{termini che dipendono solo da}$$

$$\mathcal{E}_s^{(1)}, \ldots\ldots, \mathcal{E}_s^{(n-1)} \; .$$

Risulta

$$(3.16) \qquad \mathcal{E}_{rs}^{(n)} = e_{rs}^{(n)} + l_{rs}^{(n)} \; .$$

con

$$(3.17) \qquad e_{rs}^{(n)} = \frac{1}{2} \left(u_{r,s}^{(n)} + u_{s,r}^{(n)} \right)$$

e $l_{rs}^{(n)}$ dipendente solo dalle derivate di $u_s^{(1)}, \ldots\ldots, u_s^{(n-1)}$.

Da quanto si è detto segue che l'uguaglianza

$$(3.18) \qquad Y_r^{(n)} = -\left(\frac{\partial W}{\partial \mathcal{E}_r}\right)^{(n)}$$

ha come conseguenza che

$$(3.19) \qquad Y_r^{(n)} = \eta_r^{(n)} + p_r^{(n)} \qquad ,$$

ove $p_r^{(n)}$ dipende solo dalle derivate di $u_s^{(1)}, \ldots\ldots, u_s^{(n-1)}$ ed è nullo per $n=1$ mentre

$$(3.20) \qquad \eta_r^{(n)} = -\sum_{s=1}^{6} M_{rs} \, e_s^{(n)} = -\frac{\partial W^{(2)}(e^{(n)})}{\partial e_r^{(n)}} \qquad .$$

G. Grioli

Da (3.19), (3.20) si riconosce che la dipendenza di $Y_r^{(n)}$ dalle $e_1^{(n)}$ è proprio la stessa che nel caso lineare.

Nel caso del problema misto, dal sistema (3.3), (3.4), per derivazione rispetto a θ per $\theta = 0$, si deduce[10]

$$(3.20) \qquad Y_{rs}^{(n),\,s} = k^* F_r^{(n)} \,, \qquad\qquad (\text{ in } C^*)$$

$$(3.21) \quad Y_{rs}^{(n)} N_*^s = \begin{cases} f_r^{(n)} & \text{su } \sum_1^* \\[2em] f_r^{(n)} + \phi_r^{(n)} & \text{su } \sum_2^* \,, \end{cases}$$

ove $k^* F_r^{(1)}, f_r^{(1)}$ coincidono con le forze assegnate e $\phi_r^{(1)}$ con la reazione[4], mentre per $n > 1$ si deve ritenere $\left[1,\ \text{pag. } 34\right]$.

$$(3.22) \quad \begin{cases} k^* F_r^{(n)} = - \displaystyle\sum_{q=1}^{n-1} \binom{n}{q} \left[Y_{lm}^{(q)} u_r^{(n-q),\,m} \right]^{,\,1} \,, \\[2.5em] f_r^{(n)} = - \displaystyle\sum_{q=1}^{n-1} \binom{n}{q} Y_{lm}^{(q)} u_r^{(n-q),\,m} N_*^1 \,. \end{cases} \qquad (n=2,\ldots),$$

---

(10) Si suppone che su $\sum_1^*$ siano assegnate la forze, mentre su $\sum_2^*$ sono presenti dei vincoli.

G. Grioli

Le (3.22) sono formate con le soluzioni dei sistemi successivi del tipo (3.20), (3.21) corrispondenti ai primi $n - 1$ termini degli sviluppi (3.9), (3.10) ed è interessante osservare che condizioni d'integrabilità del sistema d'indice n agenti sui secondi membri delle (3.20), (3.21) impongono ad $u_1, \ldots u_{n-1}$, condizioni integrali che possono non essere verificate in casi eccezionali.. In particolare,la teoria lineare, se considerata come primo termine di uno sviluppo del tipo (3.9) perde di significato quando ciò accade[11]. Tali casi eccezionali, facilmente individuabili, non si presentano in assenza di assi di equilibrio.

E' ormai chiaro che la determinazione del termine di ordine n della soluzione può farsi con il metodo esposto nella lettura II se per il sistema (3.20), (3.21) associato alle (3.19) sussiste un teorema del tipo di quello di Menabrea.

Ciò effettivamente è, come si riconosce effettuando sulle (3.20), (3.21) un'opportuna trasformazione. Basterà osservare che, in base a (3.19), le (3.20), (3.21) possono porsi nella forma

$$(3.22) \qquad \eta_{rs}^{(n),s} = k^* \psi_r^{(n)}, \qquad\qquad (\text{in } C^*)$$

$$(3.23) \qquad \eta_{rs}^{(n)} N_*^s = \begin{cases} \varphi_r^{(n)} & (\text{ su } \Sigma_1^*) \\ \\ \varphi_r^{(n)} + \phi_r^{(n)}, & (\text{ su } \Sigma_2^*), \end{cases}$$

---

(11) Almeno come teoria statica $\begin{bmatrix} 7, & 8, & 9, & 10 \end{bmatrix}$.

con

$$\begin{cases} k^* \psi_r^{(n)} = k^* F_r^{(n)} - p_{rs}^{(n)}\overset{\ast}{s}{}^{s} \;, \\[2em] \varphi_r^{(n)} = f_r^{(n)} - p_{rs}^{(n)} \overset{\ast}{N}{}^{s} \end{cases}$$

(3.24)

I secondi membri di (3.22), (3.23) sono espressi mediante le soluzioni $u_r^{(1)}, \ldots\ldots\ldots, u_r^{(n-1)}$ dei primi $n-1$ sistemi successivi, mentre $\eta_{rs}^{(n)}$ dipende soltanto dalle $e_{rs}^{(n)}$ e nel modo stesso con cui lo stress è legato allo strain nel caso lineare. Si può, anzi, scrivere

$$(3.25) \qquad e_r^{(n)} = - \frac{\partial W^{(2)}(\eta)^{(n)}}{\partial \eta_r^{(n)}}$$

con

$$(3.26) \qquad W^{(2)}(\eta^{(n)}) = \frac{1}{2} \sum_{r,s=1}^{6} m_{rs}\, \eta_r^{(n)} \eta_s^{(n)} > 0 \;,$$

La W esprime evidentemente la densità di energia potenziale elastica - in una teoria lineare - corrispondente allo stress $\eta_{rs}^{(n)}$. E' chiaro ormai che il sistema (3.22), (3.23) è analogo a quello del caso lineare con forze assegnate $k^* \psi_r^{(n)}, \varphi_r^{(n)}$, reazioni $\phi_r^{(n)}$ e caratteristiche di deformazioni espresse dalle $e_r^{(n)}$ e legate alle $\eta_r^{(n)}$ dalla (3.25) . Ciò basta per assicurare la validità del teorema di Menabrea per lo stress $\eta_r^{(n)}$.

<u>Osservazione</u>. Nelle applicazioni concrete lo sviluppo (3.9,1) va fermato ad un certo termine di ordine p nel senso che si ritiene trascurabile il contributo dei termini successivi. A tale proposito è opportuno fare una pre-

G. Grioli

cisazione, ad evitare equivoci. Precisamente si dovrà ritenere che nello sviluppo (3.9) è lecito fermarsi al termine di grado p se e solo se sono trascurabili tutti i prodotti

$$(3.27) \qquad u_{r,s}^{(1)\mu_1} \, u_{t,n}^{(2)\mu_2} \, \ldots\ldots\ldots \, u_{\gamma,\sigma}^{(q)\mu_q} \, q \, \frac{\theta^q}{q!}$$

con $\mu_1, \mu_2, \ldots\ldots, \mu_q$ interi positivi o nulli e soddisfacenti alla condizione

$$(3.28) \qquad \mu_1 + 2\mu_2 + \ldots\ldots\ldots + q\mu_q = q \geqslant p + 1.$$

Ciò implica la trascurabilità di $\mathcal{E}_r^{(q)} \frac{\theta^q}{q!}$ per $q > p$. Conseguentemente lo sviluppo (3.10) va arrestato al termine di grado p. Ciò si giustifica ammettendo la naturale ipotesi che le derivate della $W(\mathcal{E})$ rispetto alle $\mathcal{E}_{rs}$, valutate per $\theta = 0$ e di ordine maggiore di p abbiano grandezza comparabile o trascurabile rispetto alle derivate seconde (per $\theta = 0$) il che è quanto dire che i coefficienti elastici della teoria lineare (coefficienti di Lamè nel caso lineare) non sono trascurabili rispetto a quelli della teoria non lineare. Sotto tale ipotesi, la trascurabilità dei prodotti (3.27) per $q > p$ implica, in base a (3.19), (3.25), quella dei termini $Y_r^{(q)} \frac{\theta^q}{q!}$ nello sviluppo (3.10).

G. Grioli

BIBLIOGRAFIA

[1] G. Grioli ''Mathematical Theory of Elastic Equilibrium (Recent Results)'' Springer-Verlag. Berlin Gottingen Heidelberg (1962).

[2] F. Stoppelli ''Un teorema di esistenza ed unicità relativo alle equazioni dell'elastostatica isoterma per le deformazioni finite'' Ricerche Mat. 3, 247-267 (1954).

[3] F. Stoppelli ''Sulla sviluppabilità in serie di potenze di un parametro delle soluzioni delle equazioni dell'Elastostatica isoterma'' Ricerche Mat. 4, 58-73 (1955).

[4] F. Stoppelli ''Sull'esistenza di soluzioni delle equazioni dell'Elastostatica isoterma nel caso di sollecitazioni dotate di assi di equilibrio'' Memoria I; Ricerche di Mat. 6, 241-287 (1957).

[5] F. Stoppelli : Titolo come 4 Memoria II, 7, 71-101 (1958).

[6] F. Stoppelli : Titolo come 4 Memoria III, 7, 138-152 (1958).

[7] G. Capriz ''Sopra le deformazioni elastiche finite di un solido tubolare'' Rend. Mat. Roma; 15, 228-262 (1956).

[8] G. Capriz ''Alcune osservazioni su problemi di instabilità delle travi elastiche'' Rend. Mat. Roma; 16, 23-42 (1957)

[9] G. Capriz ''Alcune osservazioni sulla instabilità di una trave sollecitata a torsione'' Riv. Mat. Univ. Parma, 8, 145-160 (1957)

G. Grioli

[10] G. Capriz ''Sui casi di ''incompatibilità'' tra l'elastostatica classica e la teoria delle deformazioni elastiche finite'' Riv. Mat. Univ. Parma, 10, 119-129 (1959).